Next-Generation Machine Learning with Spark

Covers XGBoost, LightGBM, Spark NLP, Distributed Deep Learning with Keras, and More

Butch Quinto

Apress®

Next-Generation Machine Learning with Spark: Covers XGBoost, LightGBM, Spark NLP, Distributed Deep Learning with Keras, and More

Butch Quinto
Carson, CA, USA

ISBN-13 (pbk): 978-1-4842-5668-8 ISBN-13 (electronic): 978-1-4842-5669-5
https://doi.org/10.1007/978-1-4842-5669-5

Managing Director, Apress Media LLC: Welmoed Spahr
Acquisitions Editor: Susan McDermott
Development Editor: Laura Berendson
Coordinating Editor: Rita Fernando

Cover designed by eStudioCalamar

Cover image designed by Freepik (www.freepik.com)

Distributed to the book trade worldwide by Springer Science+Business Media New York, 1 New York Plaza, New York, NY 10004. Phone 1-800-SPRINGER, fax (201) 348-4505, e-mail orders-ny@springer-sbm.com, or visit www.springeronline.com. Apress Media, LLC is a California LLC and the sole member (owner) is Springer Science + Business Media Finance Inc (SSBM Finance Inc). SSBM Finance Inc is a **Delaware** corporation.

For information on translations, please e-mail rights@apress.com, or visit http://www.apress.com/rights-permissions.

Apress titles may be purchased in bulk for academic, corporate, or promotional use. eBook versions and licenses are also available for most titles. For more information, reference our Print and eBook Bulk Sales web page at http://www.apress.com/bulk-sales.

Any source code or other supplementary material referenced by the author in this book is available to readers on GitHub via the book's product page, located at www.apress.com/9781484256688. For more detailed information, please visit http://www.apress.com/source-code.

Printed on acid-free paper

This book is dedicated to my wife Aileen;
my children, Matthew, Timothy, and Olivia;
my sisters, Kat and Kristel;
and my parents, Edgar and Cynthia.

Table of Contents

About the Author

Butch Quinto is Founder and Chief AI Officer at Intelvi AI, an artificial intelligence company that develops cutting-edge solutions for the defense, industrial, and transportation industries. As Chief AI Officer, Butch heads strategy, innovation, research, and development. Previously, he was the Director of Artificial Intelligence at a leading technology firm and Chief Data Officer at an AI start-up. As Director of Analytics at Deloitte, he led the development of several enterprise-grade AI and IoT solutions as well as strategy, business development, and venture capital due diligence. Butch has more than 20 years of experience in various technology and leadership roles in several industries including banking and finance, telecommunications, government, utilities, transportation, e-commerce, retail, manufacturing, and bioinformatics. He is also the author of *Next-Generation Big Data* (Apress, 2018) and a member of the Association for the Advancement of Artificial Intelligence and the American Association for the Advancement of Science.

About the Technical Reviewer

Irfan Elahi has years of multidisciplinary experience in data science and machine learning. He has worked in a number of verticals such as consultancy firms, his own start-ups, and academia research lab. Over the years he has worked on a number of data science and machine learning projects in different niches such as telecommunication, retail, Web, public sector, and energy with the goal to enable businesses to derive immense value from their data assets.

Acknowledgments

I would like to thank everyone at Apress, particularly Rita Fernando Kim, Laura C. Berendson, and Susan McDermott for all the help and support in getting this book published. It was a pleasure working with the Apress team. Several people have contributed to this book directly and indirectly. Thanks to Matei Zaharia, Joeri Hermans, Max Pumperla, Fangzhou Yang, Alejandro Correa Bahnsen, Zygmunt Zawadzki, and Irfan Elahi. Thanks to Databricks and the entire Apache Spark, ML, and AI community. A special acknowledgment to Kat, Kristel, Edgar, and Cynthia for the encouragement and support. Last but not least, thanks to my wife Aileen and children Matthew, Timothy, and Olivia.

Introduction

This book provides an accessible introduction to Spark and Spark MLlib. However, this is not yet another book on the standard Spark MLlib algorithms. This book focuses on powerful third-party machine learning algorithms and libraries beyond what is available in the standard Spark MLlib library. Some of the advanced topics I cover include XGBoost4J-Spark, LightGBM on Spark, Isolation Forest, Spark NLP, Stanford CoreNLP, Alluxio, Distributed Deep Learning with Keras and Spark using Elephas and Distributed Keras, and more.

I assume no prior experience with Spark and Spark MLlib. However, some knowledge of machine learning, Scala, and Python is helpful if you want to follow the examples in this book. I highly recommend you work through the examples and experiment with the code samples to get the most out of this book. Chapter 1 starts off with a quick introduction to machine learning. Chapter 2 provides an introduction to Spark and Spark MLlib. If you want to jump right into more advanced topics, feel free to go straight to the chapter that interests you. This book is for practitioners. I tried to keep the book as simple and practical as possible, focusing on a hands-on approach rather than concentrating on theory (even though there is also plenty of that in this book). If you need a more thorough introduction to machine learning, I suggest you use a companion reference such as *An Introduction to Statistical Learning* by Gareth James, Daniela Witten, Trevor Hastie, and Robert Tibshirani (Springer, 2017) and *The Elements of Statistical Learning* by Trevor Hastie, Robert Tibshirani, and Jerome Friedman (Springer, 2016). For more information on Spark MLlib, consult Apache Spark's *Machine Learning Library (MLlib) Guide* online. For a thorough treatment of deep learning, I recommend *Deep Learning* by Ian Goodfellow, Yoshua Bengio, and Aaron Courville (MIT Press, 2016).

CHAPTER 1

Introduction to Machine Learning

> *I could give you the usual arguments. But the truth is that the prospect of discovery is too sweet.*
>
> —Geoffrey Hinton[i]

Machine learning (ML) is a subfield of artificial intelligence, the science and engineering of making intelligent machines.[ii] One of the pioneers of artificial intelligence, Arthur Samuel, defined machine learning as a "field of study that gives computers the ability to learn without being explicitly programmed."[iii] Figure 1-1 shows the relationship between artificial intelligence, machine learning, and deep learning. Artificial intelligence (AI) encompasses other fields, which means that while all machine learning is AI, not all AI is machine learning. Another branch of artificial intelligence, symbolic artificial intelligence, was the predominant AI research paradigm for much of the twentieth century.[iv] Symbolic artificial intelligence implementations are referred to as expert systems or knowledge graphs which are in essence rules engines that use if-then statements to draw logical conclusions using deductive reasoning. As you can imagine, symbolic AI suffers from several key limitations; chief among them is the complexity of revising rules once they are defined in the rules engine. Adding more rules increases the knowledge in the rules engine, but it cannot alter existing knowledge.[v] Machine learning models on the other hand are more flexible. They can be retrained on new data to learn something new or revise existing knowledge. Symbolic AI also involves significant human intervention. It relies on human knowledge and requires humans to hard-code the rules in the rules engine. On the other hand, machine learning is more dynamic, learning and recognizing patterns from input data to produce the desired output.

B. Quinto, *Next-Generation Machine Learning with Spark*, https://doi.org/10.1007/978-1-4842-5669-5_1

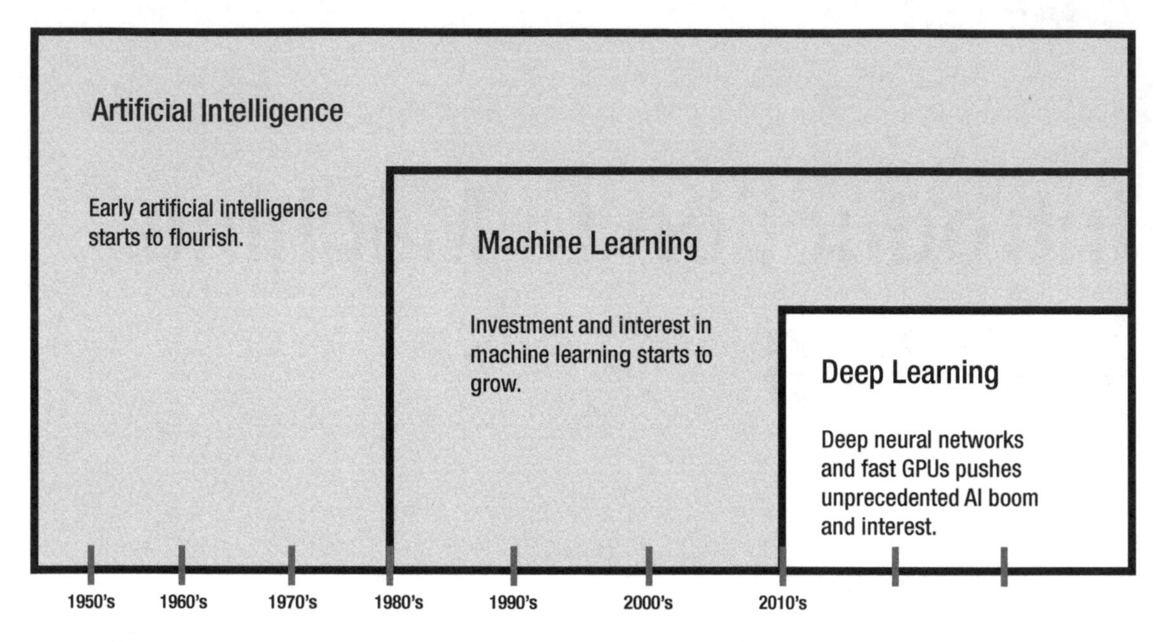

Figure 1-1. *Relationship between AI, machine learning, and deep learning*[vi]

The resurgence of deep learning in the mid-2000s put the spotlight back to the connectionist approach to artificial intelligence and machine learning in general. The resurgence of deep learning, availability of high-speed graphics processing units (GPUs), emergence of big data, and investments from companies such as Google, Facebook, Amazon, Microsoft, and IBM created the perfect storm that fueled the renaissance of artificial intelligence.

AI and Machine Learning Use Cases

The past decade has seen an astonishing series of advances in machine learning. These breakthroughs are disrupting our everyday life and making an impact in just about every vertical you can think of. This is by no means an exhaustive list of machine learning use cases, but it gives an indication of the many innovations that are happening across every industry.

Retail

Retail was one of the first industries to reap the benefits of machine learning. For years online shopping sites have relied on collaborative and content-based filtering algorithms to personalize shopping experience. Online recommendation and highly targeted marketing campaigns generate millions and sometimes billions of revenues for retailers. Perhaps the poster child for ML-powered online recommendation and personalization, Amazon is the most popular (and successful) online retailer to realize the benefits of machine learning. According to a study conducted by McKinsey, 35% of Amazon.com's revenue comes from its recommendation engine.[vii] Additional machine learning applications for retail include shelf space planning, planogram optimization, targeted marketing, customer segmentation, and demand forecasting.

Transportation

Almost every major auto manufacturer is working on AI-enabled autonomous vehicles powered by deep neural networks. These vehicles are fitted with GPU-enabled computers that can process more than 100 trillion operations per second (TOPS) for real-time AI perception, navigation, and path planning. Transportation and logistics companies such as UPS and FedEx use machine learning for route and fuel optimization, fleet monitoring, preventive maintenance, travel time estimation, and intelligent geofencing.

Financial Services

Predicting customer lifetime value (CLV), credit risk prediction, and fraud detection are some of the key machine learning use cases for financial services. Hedge funds and investment banks use machine learning to analyze data from the Twitter Firehose to detect tweets that are likely to move the market. Other common machine learning use cases for financial services include predicting next best action, churn prediction, sentiment analysis, and multichannel marketing attribution.

Healthcare and Biotechnology

Healthcare is a crucial area for artificial intelligence and machine learning research and applications. Hospitals and healthcare start-ups are using AI and machine learning to assist in accurately diagnosing life-threatening illnesses such as heart disease, cancer, and tuberculosis. AI-driven drug discovery as well as imaging and diagnostics are the most represented sector where AI is gaining traction[viii]. AI is also revolutionizing the way biotechnology and genomics research is conducted leading to new innovations in pathway analysis, microarray analysis, gene prediction, and functional annotation.[ix]

Manufacturing

Forward-thinking manufacturers are using deep learning for quality inspection to detect defects such as cracks, uneven edges, and scratches on hardware products. Survival analysis has been used for years by manufacturing and industrial engineers to predict time to failure of heavy-duty equipment. AI-powered robots are automating the manufacturing process, operating faster and with greater precision than their human counterpart resulting in increased productivity and lower product defects. The arrival of Internet of Things (IoT) and abundance of sensor data is expanding the number of machine learning applications for this industry.

Government

Machine learning has a wide range of applications in government. For instance, public utility companies have been using machine learning to monitor utility pipes. Anomaly detection techniques help detect leaks and burst pipes that can cause city-wide service interruptions and millions in property damage. Machine learning has also been employed in real-time water quality monitoring, preventing diseases from contamination and potentially saving lives. To conserve energy, publicly owned energy companies use machine learning to adjust energy output accordingly by determining peaks and troughs in electricity usage. AI-enabled cybersecurity is another fast-growing field and a critical government use case particularly in this day in age.

Machine Learning and Data

Machine learning models are built using a combination of algorithms and data. Using powerful algorithms is crucial, but equally important (and some might say more important) is training with ample amounts of high-quality data. Generally speaking, machine learning performs better with more data. The concept was first presented in 2001 by Microsoft Researchers Michele Banko and Eric Brill in their influential paper "Scaling to Very Very Large Corpora for Natural Language Disambiguation." Google's Research Director Peter Norvig further popularized the concept in his paper "The Unreasonable Effectiveness of Data."[x] However, more important than quantity is quality. Every high-quality model starts with high-quality features. This is where feature engineering enters the picture. Feature engineering is the process that transforms raw data into high-quality features. It is usually the most arduous part of the entire machine learning process, but it's also the most important. I'll discuss feature engineering in more detail later in the chapter. But for the meantime, let's take a look at a typical machine learning dataset. Figure 1-2 shows a subset of the Iris dataset. I will use this dataset in some of my examples throughout the book.

	Petal length	Petal width	Sepal length	Sepal width	Class label
1	3.2	4.8	7.6	6.4	Virginica
2	4.1	6.1	3.2	5.7	Virginica
4	5.4	5.5	4.5	3.9	Setosa
5	3.8	4.9	8.7	6.2	Versicolor
6	4.2	6.8	3.7	4.5	Setosa
7	5.7	7.3	4.6	4.1	Versicolor

Samples (observation, instances)

Features (attributes, dimensions)

Targets (class labels)

Figure 1-2. *Machine learning dataset*

Observations

A single row of data is known as an observation or instance.

Features

A feature is an attribute of an observation. The features are the independent variables used as inputs to the model. In Figure 1-2, the features are the petal length, petal width, sepal length, and sepal width.

Class Labels

The class label is the dependent variable in the dataset. It is the thing that we are trying to predict. It is the output. In our example, we are trying to predict the type of Iris flower: Virginica, Setosa, and Versicolor.

Models

A model is a mathematical construct that has predictive power. It estimates the relationship between the independent and dependent variables in a dataset.[xi]

Machine Learning Methods

There are different types of machine learning methods. Deciding which to use depends largely on what you want to accomplish as well as the kind of raw data that you have.

Supervised Learning

Supervised learning is a machine learning task that makes prediction using a training dataset. Supervised learning can be categorized into either classification or regression. Regression is for predicting continuous values such as “price”, “temperature”, or “distance”, while classification is for predicting categories such as “yes” or “no”, “spam” or “not spam”, or “malignant” or “benign”.

Classification

Classification is perhaps the most common supervised machine learning task. You most likely have already encountered applications that utilized classification without even realizing it. Popular use cases include medical diagnosis, targeted marketing, spam detection, credit risk prediction, and sentiment analysis, to mention a few. There are three types of classification tasks: binary, multiclass, and multilabel.

Binary Classification

A task is binary or binomial classification if there are only two categories. For example, when using binary classification algorithm for spam detection, the output variable can have two categories, "spam" or "not spam". For detecting cancer, the categories can be "malignant" or "benign". For targeted marketing, predicting the likelihood of someone buying an item such as milk, the categories can simply be "yes" or "no".

Multiclass Classification

Multiclass or multinomial classification tasks have three or more categories. For example, to predict weather conditions you might have five categories: "rainy", "cloudy", "sunny", "snowy" and "windy". To extend our targeted marketing example, multiclass classification can be used to predict if a customer is more likely to buy whole milk, reduced-fat milk, low-fat milk, or skim milk.

Multilabel Classification

In multilabel classification, multiple categories can be assigned to each observation. In contrast, only one category can be assigned to an observation in multiclass classification. Using our targeted marketing example, multilabel classification is used not only to predict if a customer is more likely to buy milk but other items as well such as cookies, butter, hotdogs, or bread.

Classification and Regression Algorithms

Various classification and regression algorithms have been developed over the years. They vary in approach, features, complexity, ease of use, training performance, and predictive accuracy. I describe the most popular ones in the following text. I cover them in greater detail in Chapter 3.

Support Vector Machine

Support vector machine (SVM) is a popular algorithm that works by finding the optimal hyperplane that maximizes the margin between two classes, dividing the data points into separate classes by as wide a gap as possible. The data points closest to the classification boundary are known as support vectors.

Logistic Regression

Logistic regression is a linear classifier that predicts probabilities. It uses a logistic (sigmoid) function to transform its output into a probability value that can be mapped to two (binary) classes. Multiclass classification is supported through multinomial logistic (softmax) regression.[xii] A linear classifier such as logistic regression is suitable when your data has a clear decision boundary.

Naïve Bayes

Naïve Bayes is a simple multiclass linear classification algorithm based on Bayes' theorem. Naïve Bayes got its name because it naively assumes that the features in a dataset are independent, ignoring any possible correlations between features. This is not the case in real-world scenarios, still Naïve Bayes tends to perform well especially on small datasets or datasets with high dimensionality. Like linear classifiers, it performs poorly on nonlinear classification problems. Naïve Bayes is a computationally efficient and highly scalable algorithm, only requiring a single pass to the dataset. It is a good baseline model for classification tasks using large datasets. It works by finding the probability that a point belongs to a class given a set of features.

Multilayer Perceptron

Multilayer perceptron is a feedforward artificial network that consists of several fully connected layers of nodes. Nodes in the input layer correspond to the input dataset. Nodes in the intermediate layers utilize a logistic (sigmoid) function, while nodes in the final output layer use a softmax function to support multiclass classification. The number of nodes in the output layer must match the number of classes.[xiii]

Decision Trees

A decision tree predicts the value of an output variable by learning decision rules inferred from the input variables. Visually, a decision tree looks like a tree inverted with the root node at the top. Every internal node represents a test on an attribute. Leaf nodes represent a class label, while an individual branch represents the result of a test. A decision tree is easy to interpret. In contrast with linear models like logistic regression, decision trees do not require feature scaling. It is able to handle missing features and works with both continuous and categorical features.[xiv] One-hot encoding categorical features[xv] are not required and are in fact discouraged when using decision trees and tree-based ensembles. One-hot encoding creates unbalanced trees and requires trees to grow extremely deep to achieve good predictive performance. This is especially true for high-cardinality categorical features. On the downside, decision trees are sensitive to noise in the data and have a tendency to overfit. Due to this limitation, decision trees by themselves are rarely used in real-world production environments. Nowadays, decision trees serve as the base model for more powerful ensemble algorithms such as Random Forest and Gradient-Boosted Trees.

Random Forest

Random Forest is an ensemble algorithm that uses a collection of decision trees for classification and regression. It uses a method called *bagging* (bootstrap aggregation) to reduce variance while maintaining low bias. Bagging trains individual trees from subsets of the training data. In addition to bagging, Random Forest uses another method called *feature bagging*. In contrast to bagging (using subsets of observations), feature bagging uses a subset of features (columns). Feature bagging aims to reduce the correlation between the decision trees. Without feature bagging, the individual trees will be extremely similar especially in situations where there are only a few dominant features. For classification, a majority vote of the output, or the mode, of the individual trees becomes the final prediction of the model. For regression, the average of the output of the individual trees becomes the final output (Figure 3-3). Spark trains several trees in parallel since each tree is trained independently in Random Forest.

Gradient-Boosted Trees

Gradient-Boosted Tree (GBT) is another tree-based ensemble algorithm similar to Random Forest. GBTs use a technique known as *boosting to* create a strong learner from weak learners (shallow trees). GBTs train an ensemble of decision trees sequentially[xvi] with each succeeding tree decreasing the error of the previous tree. This is done by using the residuals of the previous model to fit the next model[xvii]. This residual-correction process[xviii] is performed a set number of iterations with the number of iterations determined by cross-validation, until the residuals have been fully minimized.

XGBoost

XGBoost (eXtreme Gradient Boosting) is one of the best gradient-boosted tree implementations currently available. Released on March 27, 2014, by Tianqi Chen as a research project, XGBoost has become the dominant machine learning algorithm for classification and regression. XGBoost was designed using the general principles of gradient boosting, combining weak learners into a strong learner. But while gradient-boosted trees are built sequentially – slowly learning from data to improve its prediction in succeeding iteration, XGBoost builds trees in parallel.

XGBoost produces better prediction performance by controlling model complexity and reducing overfitting through its built-in regularization. XGBoost uses an approximate algorithm to find split points when finding the best split points for a continuous feature.[xix] The approximate splitting method uses discrete bins to bucket continuous features, significantly speeding up model training. XGBoost includes another tree growing method using a histogram-based algorithm which provides an even more efficient method of bucketing continuous features into discrete bins. But while the approximate method creates a new set of bins per iteration, the histogram-based approach reuses bins over multiple iterations. This approach allows for additional optimizations that are not achievable with the approximate method, such as the ability to cache bins and parent and sibling histogram subtraction[xx]. To optimize sorting operations, XGBoost stores sorted data in in-memory units of blocks. Sorting blocks can be efficiently distributed and performed by parallel CPU cores. XGBoost can effectively handle weighted data via its weighted quantile sketch algorithm, can efficiently handle sparse data, is cache aware, and supports out-of-core computing by utilizing disk space for large datasets so data does not have to fit in memory. XGBoost is not part of the core Spark MLlib library, but it is available as an external package.

LightGBM

For years XGBoost has been everyone's favorite go-to algorithm for classification and regression. Lately, LightGBM has emerged as the new challenger to the throne. It is a relatively new tree-based gradient boosting variant similar to XGBoost. LightGBM was released on October 17, 2016, as part of Microsoft's Distributed Machine Learning Toolkit (DMTK) project. It was designed to be fast and distributed resulting in faster training speed and low memory usage. It supports GPU and parallel learning and the ability to handle large datasets.

LightGBM has been shown in several benchmarks and experiments on public datasets to be even faster and with better accuracy than XGBoost. It has several advantages over XGBoost. LightGBM utilizes histograms to bucket continuous features into discrete bins. This provides LightGBM several performance advantages over XGBoost (which uses a pre-sort-based algorithm by default for tree learning) such as reduced memory usage, reduced cost of calculating the gain for each split, and reduced communication cost for parallel learning. LightGBM achieves additional performance boost by performing histogram subtraction on its sibling and parent to calculate a node's histogram. Benchmarks online show LightGBM is 11x to 15x faster than XGBoost (without binning) in some tasks.[xxi] LightGBM generally outperforms XGBoost in terms of accuracy by growing trees leaf-wise (best-first). There are two main strategies for training decision trees, level-wise and leaf-wise. Level-wise tree growth is the traditional way of growing decision trees for most tree-based ensembles (including XGBoost). LightGBM introduced the leaf-wise growth strategy. In contrast with level-wise growth, leaf-wise growth usually converges quicker[xxii] and achieves lower loss.[xxiii] Like XGBoost, LightGBM is not part of the core Spark MLlib library, but it is available as an external package. Most of LightGBM's features have been recently ported to XGBoost. I discuss both algorithms in more detail in Chapter 3.

Regression

Regression is a supervised machine learning task for predicting continuous numeric values. Popular use cases include sales and demand forecasting, predicting stock, home or commodity prices, and weather forecasting, to mention a few. Decision Tree, Random Forest, Gradient-Boosted Tree, XGBoost, and LightGBM can also be used for regression. I discuss regression in more detail in Chapter 3.

Linear Regression

Linear regression is used for examining linear relationships between one or more independent variable(s) and a dependent variable. The analysis of the relationship between a single independent variable and a single continuous dependent variable is known as simple linear regression. Multiple regression is an extension of simple linear regression for predicting the value of a dependent variable based on multiple independent variables.

Survival Regression

Survival regression, also known as analysis of time to death or failure time analysis, is used for predicting the time when a specific event is going to happen. The main feature that differentiates survival regression from linear regression is its ability to handle censoring, a type of missing data problem where the time to event is not known.

Unsupervised Learning

Unsupervised learning is a machine learning task that finds hidden patterns and structure in the dataset without the aid of labeled responses. Unsupervised learning is ideal when you only have access to input data and training data is unavailable or hard to obtain. Common methods include clustering, topic modeling, anomaly detection, recommendations, and principal component analysis.

Clustering

Clustering is an unsupervised machine learning task for grouping unlabeled observations that have some similarities. Popular clustering use cases include customer segmentation, fraud analysis, and anomaly detection. Clustering is also often used to generate training data for classifiers in cases where training data is scarce or unavailable.

K-Means

K-Means is one of the most popular unsupervised learning algorithms for clustering. K-Means works by randomly assigning centroids that are used as the starting point for each cluster. The algorithm iteratively assigns each data point to the nearest centroid based on the Euclidean distance. It then calculates a new centroid for each cluster by calculating the mean of all the points that are part of that cluster. The algorithm stops iterating when a predefined number of iteration is reached or every data point is assigned to its nearest centroid and there are no more reassignments that can be performed.

Topic Modeling

Topic models automatically derive the themes (or topics) in a group of documents. These topics can be used for content-based recommendations, document classification, dimensionality reduction, and featurization.

Latent Dirichlet Allocation

Latent Dirichlet Allocation (LDA) was developed in 2003 by David M. Blei, Andrew Ng, and Michael Jordan, although a similar algorithm used in population genetics was also proposed by Jonathan K. Pritchard, Matthew Stephens, and Peter Donnelly in 2000. LDA, as applied to machine learning, is based on a graphical model and is the first algorithm included Spark MLlib built on GraphX. Latent Dirichlet Allocation is widely used for topic modeling.

Anomaly Detection

Anomaly or outlier detection identifies rare observations that deviate significantly and stand out from majority of the dataset. It is frequently used in discovering fraudulent financial transactions, identifying cybersecurity threats, or performing predictive maintenance, to mention a few use cases.

Isolation Forest

Isolation Forest is a tree-based ensemble algorithm for anomaly detection that was developed by Fei Tony Liu, Kai Ming Ting, and Zhi-Hua Zhou.[xxiv] Unlike most anomaly detection techniques, Isolation Forest tries to explicitly detect actual outliers instead of identifying normal data points. Isolation Forest operates based on the fact that there are usually a small number of outliers in a dataset and are therefore prone to the process of isolation.[xxv] Isolating outliers from normal data points is efficient since it requires fewer conditions. In contrast, isolating normal data points generally involves more conditions. Similar to other tree-based ensembles, Isolation Forest is built on a collection of decision trees known as *isolation trees*, with each tree having a subset of the entire dataset. An anomaly score is computed as the average anomaly score of the trees in the forest. The anomaly score is derived from the number of conditions required to split a data point. An anomaly score close to 1 signifies an anomaly, while a score less than 0.5 signifies non-anomalous observations.

One-Class Support Vector Machines

A support vector machine (SVM) classifies data by dividing the data points into separate classes by as wide a gap as possible using the most optimal hyperplane. In one-class SVM, the model is trained on data that has only one "normal" class. Data points that are unlike the normal examples are considered anomalies.

Dimensionality Reduction

Dimensionality reduction is essential when there are a high number of features in your dataset. Machine learning use cases in the fields of genomics and industrial analytics, for instance, usually involve thousands or even millions of features. High dimensionality makes models more complex, increasing the chances of overfitting. Adding more features at a certain point will actually decrease the performance of the model. Moreover, training on high-dimensional data requires significant computing resources. These are collectively known as the curse of dimensionality. Dimensionality reduction techniques aim to overcome the curse of dimensionality.

Principal Component Analysis

Principal component analysis (PCA) is an unsupervised machine learning technique used for reducing the dimensionality of the feature space. It detects correlations between features and generates a reduced number of linearly uncorrelated features while retaining most of the variance in the original dataset. These more compact, linearly uncorrelated features are called *principal components*. The principal components are sorted in descending order of their explained variance. Other dimensionality reduction technique includes singular value decomposition (SVD) and linear discriminant analysis (LDA).

Recommendations

Providing personalized recommendations is one of the most popular applications of machine learning. Almost every major retailer, such as Amazon, Alibaba, Walmart, and Target, provides some sort of personalized recommendation based on customer behavior. Streaming services such as Netflix, Hulu, and Spotify provide movie or music recommendations based on user tastes and preferences. Collaborative filtering, content-based filtering, and association rules learning (for market basket analysis) are the most popular methods for building recommendation systems. Spark MLlib supports alternating least squares (ALS) for collaborative filtering and FP-Growth and PrefixSpan for market basket analysis.

Semi-supervised Learning

In some cases, getting access to labeled data is costly and time-consuming. In situations where labeled responses are scarce, semi-supervised learning combines both supervised and unsupervised learning techniques to make predictions. In semi-supervised learning, unlabeled data is utilized to augment labeled data to improve model accuracy.

Reinforcement Learning

Reinforcement learning tries to learn through trial and error to determine which action provides the greatest reward. Reinforcement learning has three components: the agent (the decision-maker or learner), the environment (what the agent interacts with), and actions (what the agent can perform).[xxvi] This type of learning is frequently used in gaming, navigation, and robotics.

Deep Learning

Deep learning is a subfield of machine learning and artificial intelligence that uses deep, multilayered artificial neural networks. It is responsible for many of the recent breakthroughs in artificial intelligence. While deep learning can be used for more mundane classification tasks, its true power shines when applied to more complex problems such as medical diagnosis, facial recognition, self-driving cars, fraud analysis, and intelligent voice-controlled assistants.[xxvii] In certain areas, deep learning has enabled computers to match and even outperform human abilities.

Neural Networks

Neural networks are a class of algorithms that operate like interconnected neurons of the human brain. A neural network contains multiple layers that consist of interconnected nodes. There is usually an input layer, one or more hidden layers, and an output layer. Data goes through the neural network via the input layer. The hidden layers process the data through a network of weighted connections. The nodes in the hidden layer assign weights to the input and combine it with a set of coefficients along the way. The data goes through a node's activation function which determines the output of the layer. Finally, the data reaches the output layer which produces the final output of the neural network.[xxviii] Neural networks with several hidden layers are known as "deep" neural networks. The more layers, the deeper the network, and generally the deeper the network, the more sophisticated the learning become and the more complex problems it can solve.

Convolutional Neural Networks

A convolutional neural network (convnet or CNN for short) is a type of neural network that is particularly good at analyzing images (although they can also be applied to audio and text data). The neurons in the layers of a convolutional neural network are arranged in three dimensions: height, width, and depth. CNNs use convolutional layers to learn local patterns in its input feature space (images) such as textures and edges. In contrast, fully connected (dense) layers learn global patterns.[xxix] The neurons in a convolutional layer are only connected to a small region of the layer preceding it instead of all of the neurons as is the case with dense layers. A dense layer's fully connected structure can lead to extremely large number of parameters which is inefficient and could quickly lead to overfitting.[xxx] I cover deep learning and deep convolutional neural networks in greater detail in Chapter 7.

Feature Engineering

Feature engineering is the process of transforming data to create features that can be used for training machine learning models. Oftentimes raw data needs to be transformed through several data preparation and extraction techniques. For instance, dimensionality reduction might be needed for extremely large datasets. New features might have to be created from other features. Distance-based algorithms require feature scaling. Some algorithms perform better when categorical features are one-hot encoded. Text data usually require tokenization and feature vectorization. After transforming raw data into features, they are evaluated and the ones with the most predictive power are selected.

Feature engineering is an important aspect of machine learning. It is axiomatic in almost every machine learning endeavor to generate highly relevant features if it is to succeed. Unfortunately, feature engineering is a complicated and time-consuming task that often requires domain expertise. It is an iterative process that involves brainstorming features, creating them and studying their impact on model accuracy. In fact, the typical data scientists spend most of their time preparing data according to a survey by Forbes (Figure 1-3).[xxxi]

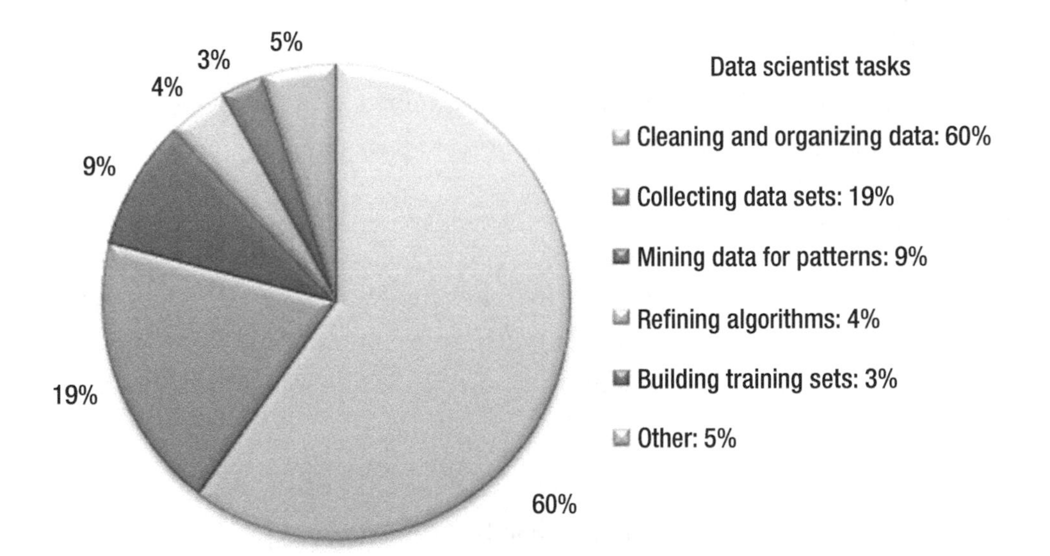

Figure 1-3. *Data preparation accounts for about 80% of the work of data scientists*

Feature engineering tasks can be divided into several categories: feature selection, feature importance, feature extraction, and feature construction.[xxxii]

Feature Selection

Feature selection is an important preprocessing step in identifying important features and eliminating irrelevant or redundant ones. It improves prediction performance and model training efficiency and reduces dimensionality. Irrelevant features must be removed since they can negatively affect the model's accuracy as well as slow down model training. Certain features may not have any predictive power or they could be redundant with other features. But how do we determine if the features are relevant or not? Domain knowledge is critical. For example, if you are building a model to predict the probability of loan default, it helps to know what factors to consider in quantifying credit risk. You might start with the borrower's debt-to-income ratio. There are other borrow-specific factors to consider such as the borrower's credit score, employment length, employment title, and marital status. Market-wide considerations such as economic growth might also be important. Demographic and psychographic information should also be considered. Once you have a list of features, there are several ways to objectively determine their importance. There are various feature selection methods to aid you in selecting the right features for your models.[xxxiii]

Filter Methods

Filter methods assign ranking to each feature using statistical techniques such as chi-squared test, correlation coefficients, and information gain.

Wrapper Methods

Wrapper methods use a subset of features to train a model. You can then add or delete features based on the performance of the model. Common examples of wrapper methods are recursive feature elimination, backward elimination, and forward selection.

Embedded Methods

Embedded methods combine techniques used by filter and wrapper methods. Popular examples include LASSO and RIDGE regression, regularized trees, and random multinomial logit.[xxxiv]

Feature Importance

Tree-based ensembles such as Random Forest, XGBoost, and LightGBM provide a feature selection method based on a feature importance score that is calculated for each feature. The higher the score, the more important the feature is toward increasing model accuracy. Feature importance in Random Forest is also known as gini-based importance or mean decrease in impurity (MDI). Some implementation of Random Forest utilizes a different method to calculate feature importance known as accuracy-based importance or mean decrease in accuracy (MDA).[xxxv] Accuracy-based importance is calculated based on decreases in prediction accuracy as features are randomly permuted. I discuss feature importance in Random Forest, XGBoost, and LightGBM in greater detail in Chapter 3.

Correlation coefficients are a rudimentary form of feature selection method. Correlation coefficients represent the strength of the linear relationship between two variables. For linear problems you can use correlation to select relevant features (feature-class correlation) and identify redundant features (intra-feature correlation).

Feature Extraction

Feature extraction is essential when there are a high number of features in your dataset. Machine learning use cases in the fields of genomics and industrial analytics, for instance, usually involve thousands or even millions of features. High dimensionality makes models more complex, increasing the chances of overfitting. Moreover, training on high-dimensional data requires significant computing resources. Feature extraction usually involves using dimensionality reduction techniques. Principal component analysis (PCA), linear discriminant analysis (LDA), and singular value decomposition (SVD) are some of the most popular dimensionality reduction algorithms that are also used for feature extraction.

Feature Construction

To improve a model's accuracy, sometimes new features need to be constructed from existing features. There are several ways to do this. You can combine or aggregate features. In some cases, you might want to split them. For instance, your model might benefit from splitting a timestamp attribute, which is common in most transactional data, into several, more granular attributes: seconds, minutes, hour, day, month, and year. Then, you might want to use these attributes to construct even more features such as dayofweek, weekofmonth, monthofyear, and so on. Feature construction is part art, part science and is one of the most difficult and time-consuming aspect of feature engineering. Mastery of feature construction is usually what differentiates a seasoned data scientist from a novice.

Model Evaluation

In classification, each data point has a known label and a model-produced predicted class. By comparing the known label and the predicted class for each data point, the results can be classified into one of the four categories: true positive (TP) where the predicted class is positive and label is positive, true negative (TN) where the predicted class is negative and label is negative, false positive (FP) where the predicted class is positive but label is negative, and false negative (FN) where the predicted class is negative and label is positive. These four values form the basis for most evaluation metrics for classification tasks. They are often presented in a table called a confusion matrix (Table 1-1).

Table 1-1. *Confusion Matrix*

	Negative (Predicted)	Positive (Predicted)
Negative (Actual)	True Negative	False Positive
Positive (Actual)	False Negative	True Positive

Accuracy

Accuracy is an evaluation metric for classification models. It is defined as the number of correct predictions divided by the total number of predictions.

$$Accuracy = \frac{True\ Positive + True\ Negative}{True\ Positive + True\ Negative + False\ Positive + False\ Positive}$$

Accuracy is not the ideal metric in situations where you have imbalanced datasets. To illustrate with an example, consider a hypothetical classification task with 90 negative and 10 positive samples; classifying all as negative gives a 0.90 accuracy score. Precision and recall are better metrics for evaluating models trained with class-imbalanced data.

Precision

Precision is defined as the number of true positives divided by the number of true positives plus the number of false positives. Precision shows how often the model is correct when its prediction is positive. For instance, if your model predicted 100 cancer occurrence but 10 of them were incorrect predictions, your model's precision is 90%. Precision is a good metric to use in situations where the costs of false positives are high.

$$Precision = \frac{True\ Positive}{True\ Positive + False\ Positive}$$

Recall

Recall is a good metric to use in situations where the cost of false negatives is high. Recall is defined as the number of true positives divided by the number of true positives plus the number of false negatives.

$$Precision = \frac{True\ Positive}{True\ Positive + False\ Negative}$$

F1 Measure

The F1 measure or F1 score is the harmonic mean or weighted average of precision and recall. It is a common performance metric for evaluating multiclass classifiers. It is also a good measure when there is an uneven class distribution. The best F1 score is 1, while the worst score is 0. A good F1 measure means that you have low false negatives and low false positives. The F1 measure is defined as the following:

$$F1\ Measure = 2\ x\ \frac{Precision\ x\ Recall}{Precision + Recall}$$

Area Under the Receiver Operating Characteristic (AUROC)

The area under the receiver operating characteristic (AUROC) is a common performance metric for evaluating binary classifiers. The receiver operating characteristic (ROC) is a graph that plots the true positive rate against the false positive rate. The area under the curve (AUC) is the area below the ROC curve. The AUC can be interpreted as the probability that the model ranks a random positive example higher than a random negative example.[xxxvi] The larger the area under the curve (the closer the AUROC is to 1.0), the better the model is performing. A model with AUROC of 0.5 is useless since its predictive accuracy is just as good as random guessing.

Overfitting and Underfitting

A model's poor performance is caused by either overfitting or underfitting. Overfitting refers to a model that fits training data too well. An overfitted model performs well against training data, but performs poorly against new, unseen data. The opposite of overfitting is underfitting. With underfitting, models are too simple and have not learned the relevant patterns in the training dataset, either because the model is overregularized or needs to be trained longer. The ability of the model to fit well to new, unseen data is known as generalization. This is the goal of every model tuning exercise. Several established methods to prevent overfitting include using more data or subset of features, cross-validation, dropout, pruning, early stopping, and regularization.[xxxvii] For deep learning, data augmentation is a common form of regularization. To reduce underfitting, adding more relevant features is a recommended option. For deep learning, consider adding more nodes to a layer or adding more layers to the neural network to increase the capacity of the model.[xxxviii]

Model Selection

Model selection involves evaluating fitted machine learning models and outputs the best one by trying to fit the underlying estimator with user-specified combinations of hyperparameters. With Spark MLlib, model selection is performed with the CrossValidator and TrainValidationSplit estimators. A CrossValidator performs k-fold cross-validation and grid search for hyperparameter tuning and model selection. It splits the dataset into a set of random, nonoverlapping partitioned folds which are utilized as training and test datasets. For instance, if k=3 folds, k-fold cross-validation will generate 3 training and test dataset pairs (each fold is used as the test dataset only once), each of which uses 2/3 for training data and 1/3 for test.[xxxix] TrainValidationSplit is another estimator for hyperparameter tuning. In contrast with k-fold cross-validation (which is an expensive operation), TrainValidationSplit only evaluates each combination of parameters once, instead of k times.

Summary

This chapter serves as a quick introduction to machine learning. For a more thorough treatment, I suggest *Elements of Statistical Learning, 2nd ed.,* by Trevor Hastie, et al. (Springer, 2016) and *An Introduction to Statistical Learning* by Gareth James, et al. (Springer, 2013). For an introduction to deep learning, I recommend *Deep Learning* by Ian Goodfellow, et al. (MIT Press, 2016). While machine learning has been around for a long time, using big data to train machine learning models is a fairly recent development. As the most popular big data framework, Spark is uniquely positioned to be the preeminent platform for building large-scale, enterprise-grade machine learning applications. Let's dive into Spark and Spark MLlib in the next chapter.

References

i. Raffi Khatchadourian; "THE DOOMSDAY INVENTION," newyorker.com, 2015, `www.newyorker.com/magazine/2015/11/23/doomsday-invention-artificial-intelligence-nick-bostrom`

ii. John McCarthy; "What is artificial intelligence?", stanford.edu, 2007, `www-formal.stanford.edu/jmc/whatisai/node1.html`

iii. Chris Nicholson; "Artificial Intelligence (AI) vs. Machine Learning vs. Deep Learning," skimind.ai, 2019, `https://skymind.ai/wiki/ai-vs-machine-learning-vs-deep-learning`

iv. Marta Garnelo and Murray Shanahan; "Reconciling deep learning with symbolic artificial intelligence: representing objects and relations," sciencedirect.com, 2019, `www.sciencedirect.com/science/article/pii/S2352154618301943`

v. Chris Nicholson; "Symbolic Reasoning (Symbolic AI) and Machine Learning"; skymind.ai, 2019, `https://skymind.ai/wiki/symbolic-reasoning`

vi. Michael Copeland; "What's the Difference Between Artificial Intelligence, Machine Learning, and Deep Learning?", nvidia.com, 2016, https://blogs.nvidia.com/blog/2016/07/29/whats-difference-artificial-intelligence-machine-learning-deep-learning-ai/

vii. Ian MacKenzie, et al.; "How retailers can keep up with consumers," mckinsey.com, 2013, www.mckinsey.com/industries/retail/our-insights/how-retailers-can-keep-up-with-consumers

viii. CB Insights; "From Drug R&D To Diagnostics: 90+ Artificial Intelligence Startups In Healthcare," cbinsights.com, 2019, www.cbinsights.com/research/artificial-intelligence-startups-healthcare/

ix. Ragothaman Yennamali; "The Applications of Machine Learning in Biology," kolabtree.com, 2019, www.kolabtree.com/blog/applications-of-machine-learning-in-biology/

x. Xavier Amatriain; "In Machine Learning, What is Better: More Data or better Algorithms," kdnuggets.com, 2015, www.kdnuggets.com/2015/06/machine-learning-more-data-better-algorithms.html

xi. Mohammed Guller; "Big Data Analytics with Spark," Apress, 2015

xii. Apache Spark; "Multinomial logistic regression," spark.apache.org, 2019, https://spark.apache.org/docs/latest/ml-classification-regression.html#multinomial-logistic-regression

xiii. Apache Spark; "Multilayer perceptron classifier," spark.apache.org, 2019, https://spark.apache.org/docs/latest/ml-classification-regression.html#multilayer-perceptron-classifier

xiv. Analytics Vidhya Content Team; "A Complete Tutorial on Tree Based Modeling from Scratch (in R & Python)," AnalyticsVidhya.com, 2016, www.analyticsvidhya.com/blog/2016/04/complete-tutorial-tree-based-modeling-scratch-in-python/#one

xv. LightGBM; "Optimal Split for Categorical Features," lightgbm.readthedocs.io, 2019, https://lightgbm.readthedocs.io/en/latest/Features.html

xvi. Joseph Bradley and Manish Amde; "Random Forests and Boosting in MLlib," Databricks, 2015, https://databricks.com/blog/2015/01/21/random-forests-and-boosting-in-mllib.html

xvii. Analytics Vidhya Content Team; "An End-to-End Guide to Understand the Math behind XGBoost," analyticsvidhya.com, 2018, www.analyticsvidhya.com/blog/2018/09/an-end-to-end-guide-to-understand-the-math-behind-xgboost/

xviii. Ben Gorman; "A Kaggle Master Explains Gradient Boosting," Kaggle.com, 2017, http://blog.kaggle.com/2017/01/23/a-kaggle-master-explains-gradient-boosting/

xix. Reena Shaw; "XGBoost: A Concise Technical Overview," KDNuggets, 2017, www.kdnuggets.com/2017/10/xgboost-concise-technical-overview.html

xx. Philip Hyunsu Cho; "Fast Histogram Optimized Grower, 8x to 10x Speedup," DMLC, 2017, https://github.com/dmlc/xgboost/issues/1950

xxi. Laurae; "Benchmarking LightGBM: how fast is LightGBM vs xgboost?", medium.com, 2017, https://medium.com/implodinggradients/benchmarking-lightgbm-how-fast-is-lightgbm-vs-xgboost-15d224568031

xxii. LightGBM; "Optimization in Speed and Memory Usage," lightgbm.readthedocs.io, 2019, https://lightgbm.readthedocs.io/en/latest/Features.html

xxiii. David Marx; "Decision trees: leaf-wise (best-first) and level-wise tree traverse," stackexchange.com, 2018, https://datascience.stackexchange.com/questions/26699/decision-trees-leaf-wise-best-first-and-level-wise-tree-traverse

xxiv. Fei Tony Liu, Kai Ming Ting, Zhia-Hua Zhou; "Isolation Forest," acm.org, 2008, `https://dl.acm.org/citation.cfm?id=1511387`

xxv. Alejandro Correa Bahnsen; "Benefits of Anomaly Detection Using Isolation Forests," easysol.net, 2016, `https://blog.easysol.net/using-isolation-forests-anamoly-detection/`

xxvi. SAS; "Machine Learning," sas.com, 2019, `www.sas.com/en_us/insights/analytics/machine-learning.html`

xxvii. NVIDIA; "Deep Learning," developer.nvidia.com, 2019, `https://developer.nvidia.com/deep-learning`

xxviii. SAS; "How neural networks work," sas.com, 2019, `www.sas.com/en_us/insights/analytics/neural-networks.html`

xxix. Francois Chollet; "Deep learning for computer vision," 2018, Deep Learning with Python

xxx. Andrej Karpathy; "Convolutional Neural Networks (CNNs / ConvNets)," github.io, 2019, `http://cs231n.github.io/convolutional-networks/`

xxxi. Gil Press; "Cleaning Big Data: Most Time-Consuming, Least Enjoyable Data Science Task, Survey Says," forbes.com, 2016, `www.forbes.com/sites/gilpress/2016/03/23/data-preparation-most-time-consuming-least-enjoyable-data-science-task-survey-says/#680347536f63`

xxxii. Jason Brownlee; "Discover Feature Engineering, How to Engineer Features and How to Get Good at It," machinelearningmastery.com, 2014, `https://machinelearningmastery.com/discover-feature-engineering-how-to-engineer-features-and-how-to-get-good-at-it/`

xxxiii. Jason Brownlee; "An Introduction to Feature Selection," MachineLearningMastery.com, 2014, `https://machinelearningmastery.com/an-introduction-to-feature-selection/`

xxxiv. Saurav Kaushik; "Introduction to Feature Selection methods with an example," Analyticsvidhya.com, 2016, www.analyticsvidhya.com/blog/2016/12/introduction-to-feature-selection-methods-with-an-example-or-how-to-select-the-right-variables/

xxxv. Jake Hoare; "How is Variable Importance Calculated for a Random Forest," DisplayR, 2018, www.displayr.com/how-is-variable-importance-calculated-for-a-random-forest/

xxxvi. Google; "Classification: ROC Curve and AUC," developers.google.com, 2019, https://developers.google.com/machine-learning/crash-course/classification/roc-and-auc

xxxvii. Wayne Thompson; "Machine learning best practices: Understanding generalization," blogs.sas.com, 2017, https://blogs.sas.com/content/subconsciousmusings/2017/09/05/machine-learning-best-practices-understanding-generalization/

xxxviii. Jason Brownlee; "How to Avoid Overfitting in Deep Learning Neural Networks," machinelearningmaster.com, 2018, https://machinelearningmastery.com/introduction-to-regularization-to-reduce-overfitting-and-improve-generalization-error/

xxxix. Spark; "CrossValidator," spark.apache.org, 2019, https://spark.apache.org/docs/latest/api/scala/index.html#org.apache.spark.ml.tuning.CrossValidator

CHAPTER 2

Introduction to Spark and Spark MLlib

> *Simple models and a lot of data trump more elaborate models based on less data.*
>
> —Peter Norvig[i]

Spark is a unified big data processing framework for processing and analyzing large datasets. Spark provides high-level APIs in Scala, Python, Java, and R with powerful libraries including MLlib for machine learning, Spark SQL for SQL support, Spark Streaming for real-time streaming, and GraphX for graph processing.[ii] Spark was founded by Matei Zaharia at the University of California, Berkeley's AMPLab and was later donated to the Apache Software Foundation, becoming a top-level project on February 24, 2014.[iii] The first version was released on May 30, 2017.[iv]

Overview

Spark was developed to address the limitations of MapReduce, Hadoop's original data processing framework. Matei Zaharia saw MapReduce's limitations at UC Berkeley and Facebook (where he did his internship) and sought to create a faster and more generalized, multipurpose data processing framework that can handle iterative and interactive applications.[v] It provides a unified platform (Figure 2-1) that supports multiple types of workloads such as streaming, interactive, graph processing, machine learning, and batch.[vi] Spark jobs can run multitude of times faster than equivalent MapReduce jobs due to its fast in-memory capabilities and advanced DAG (directed acyclic graph) execution engine. Spark was written in Scala and consequently it is the de facto programming interface for

B. Quinto, *Next-Generation Machine Learning with Spark*, https://doi.org/10.1007/978-1-4842-5669-5_2

Spark. We will use Scala throughout this book. We will use PySpark, the Python API for Spark, in Chapter 7 for distributed deep learning. This chapter is an updated version of Chapter 5 from my previous book *Next-Generation Big Data* (Apress, 2018).

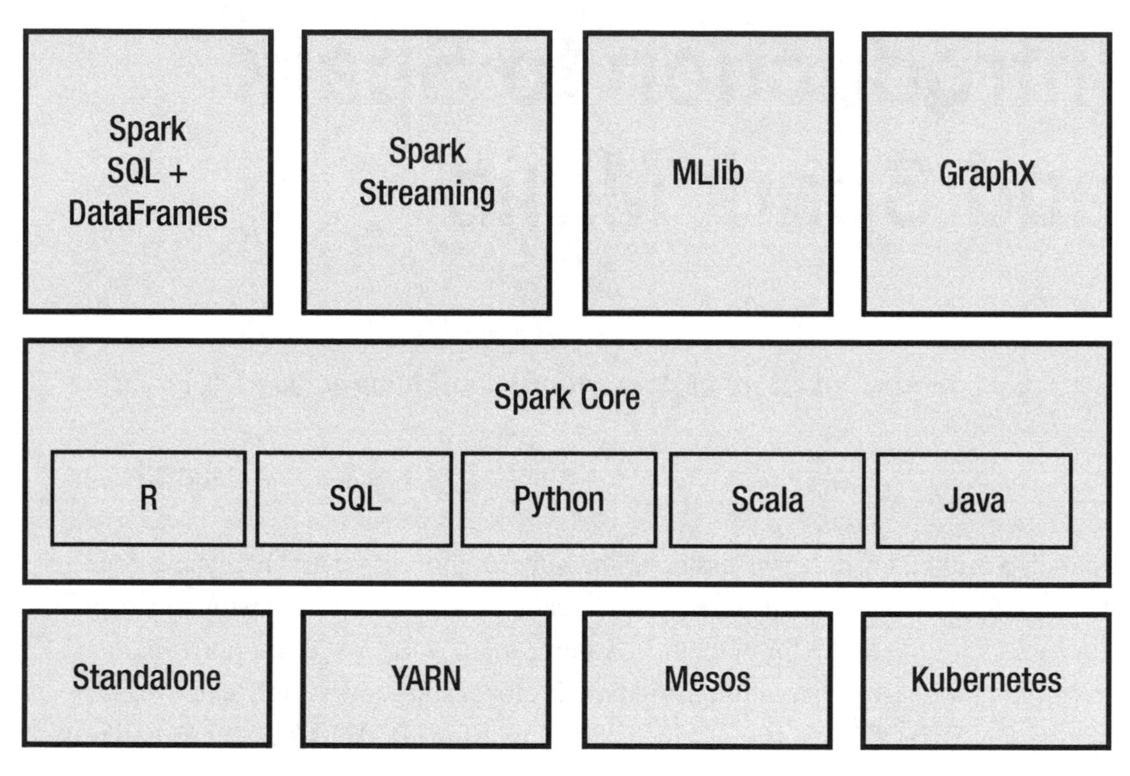

Figure 2-1. *Apache Spark ecosystem*

Cluster Managers

Cluster managers manage and allocate cluster resources. Spark supports the standalone cluster manager that comes with Spark (Standalone Scheduler), YARN, Mesos, and Kubernetes.

Architecture

At a high level, Spark distributes the execution of Spark applications' tasks across the cluster nodes (Figure 2-2). Every Spark application has a SparkContext object within its driver program. The SparkContext represents a connection to your cluster manager, which provides computing resources to your Spark applications. After connecting to the cluster, Spark acquires executors on your worker nodes. Spark then sends your application code to the executors. An application will usually run one or more jobs in response to a Spark action. Each job is then divided by Spark into smaller directed acyclic graph (DAG) of stages or tasks. Each task is then distributed and sent to executors across the worker nodes for execution.

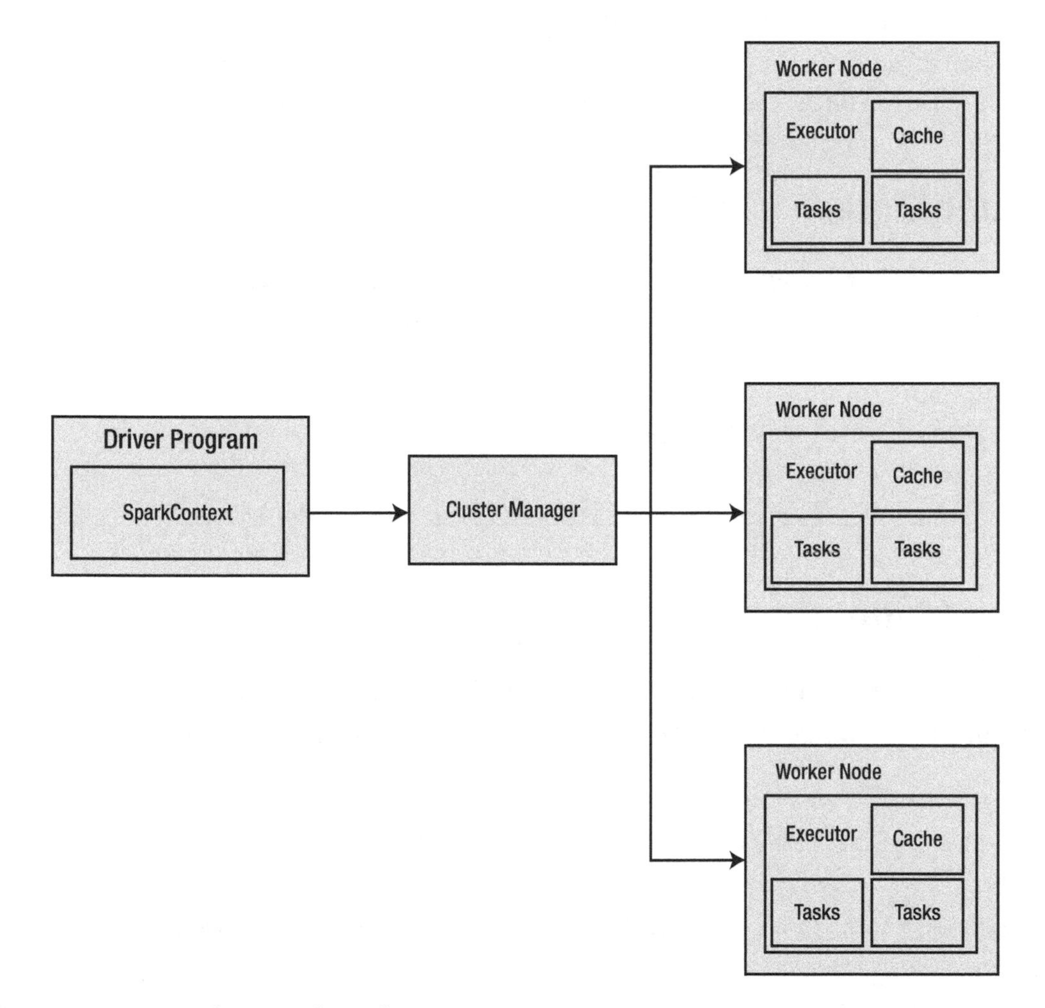

Figure 2-2. *Apache Spark architecture*

Each Spark application gets its own set of executors. Because tasks from different applications run in different JVMs, a Spark application cannot interfere with another Spark application. This also means that it's difficult for Spark applications to share data without using an external data source such as HDFS or S3. Using an off-heap memory storage such as Tachyon (a.k.a. Alluxio) can make data sharing faster and easier. I discuss Alluxio in greater detail later in the chapter.

Executing Spark Applications

You use an interactive shell (spark-shell or pyspark) or submit an application (spark-submit) to execute Spark applications. Some prefer to use interactive web-based notebooks such as Apache Zeppelin and Jupyter to interact with Spark. Commercial vendors such as Databricks and Cloudera provide their own interactive notebook environment as well. I will use the spark-shell throughout the chapter. There are two deploy modes for launching Spark applications in an environment with a cluster manager such as YARN.

Cluster Mode

In cluster mode, the driver program runs inside an application master managed by YARN. The client can exit without affecting the execution of the application. To launch applications or the spark-shell in cluster mode:

```
spark-shell --master yarn --deploy-mode cluster

spark-submit --class mypath.myClass --master yarn --deploy-mode cluster
```

Client Mode

In client mode, the driver program runs in the client. The application master is only used for requesting resources from YARN. To launch applications or the spark-shell in client mode:

```
spark-shell --master yarn --deploy-mode client

spark-submit --class mypath.myClass --master yarn --deploy-mode client
```

Introduction to the spark-shell

You typically use an interactive shell for ad hoc data analysis or exploration. It's also a good tool to learn the Spark API. Spark's interactive shell is available in Spark or Python. In the following example, we'll create an RDD of cities and convert them all to uppercase. A SparkSession named "spark" is automatically created when you start spark-shell as shown in Listing 2-1.

Listing 2-1. Introduction to spark-shell

```
spark-shell

Spark context Web UI available at http://10.0.2.15:4041
Spark context available as 'sc' (master = local[*], app id =
local-1574144576837).
Spark session available as 'spark'.
Welcome to
      ____              __
     / __/__  ___ _____/ /__
    _\ \/ _ \/ _ `/ __/  '_/
   /___/ .__/\_,_/_/ /_/\_\   version 2.4.4
      /_/

Using Scala version 2.11.12 (OpenJDK 64-Bit Server VM, Java 1.8.0_212)
Type in expressions to have them evaluated.
Type :help for more information.

scala>val myCities = sc.parallelize(List(
                                    "tokyo",
                                    "new york",
                                    "sydney",
                                    "san francisco"))

scala>val uCities = myCities.map {x =>x.toUpperCase}

scala>uCities.collect.foreach(println)
TOKYO
NEW YORK
SYDNEY
SAN FRANCISCO
```

SparkSession

As you can see in Figure 2-2, SparkContext enables access to all Spark features and capabilities. The driver program uses SparkContext to access other contexts such as StreamingContext, SQLContext, and HiveContext. Starting in Spark 2.0, SparkSession provides a single point of entry to interact with Spark. All features available through SparkContext, SQLContext, HiveContext, and StreamingContext in Spark 1.x are now accessible via SparkSession.[vii] You may still encounter code written in Spark 1.x. In Spark 1.x you would write something like this.

```
val sparkConf = new SparkConf().setAppName("MyApp").setMaster("local")

val sc = new SparkContext(sparkConf).set("spark.executor.cores", "4")

val sqlContext = new org.apache.spark.sql.SQLContext(sc)
```

In Spark 2.x, you don't have to explicitly create SparkConf, SparkContext, or SQLContext since all of their functionalities are already included in SparkSession.

```
val spark = SparkSession.
builder().
appName("MyApp").
config("spark.executor.cores", "4").
getOrCreate()
```

Resilient Distributed Dataset (RDD)

An RDD is a resilient immutable distributed collection of objects partitioned across one or more nodes in your cluster. RDDs can be processed and operated in parallel by two types of operations: transformations and actions.

Note The RDD was Spark's primary programming interface in Spark 1.x. Dataset has replaced RDD as the main API starting in Spark 2.0. Users are recommended to switch from RDD to Dataset/DataFrame due to richer programming interface and better performance. I discuss Dataset and DataFrame later in the chapter.

Creating an RDD

Creating an RDD is straightforward. You can create an RDD from an existing Scala collection or from reading from an external file stored in HDFS or S3.

parallelize

Parallelize creates an RDD from a Scala collection.

```
val data = (1 to 5).toList
val rdd = sc.parallelize(data)
val cities = sc.parallelize(List("tokyo","new york","sydney","san
francisco"))
```

textFile

TextFile creates an RDD from a text file stored in HDFS or S3.

```
val rdd = sc.textFile("hdfs://master01:9000/files/mydirectory")

val rdd = sc.textFile("s3a://mybucket/files/mydata.csv")
```

Note that an RDD is immutable. Data transformations produce another RDD instead of modifying the current RDD. RDD operations can be classified into two categories: transformation and action.

Transformations

A transformation is an operation that creates a new RDD. I describe some of the most common transformations. Refer to the online Spark documentation for a complete list.

Map

Map executes a function against each element in the RDD. It creates and returns a new RDD of the result. Map's return type doesn't necessarily have to be the same type of the original RDD.

```
val cities = sc.parallelize(List("tokyo","new york","paris","san francisco"))
val upperCaseCities = myCities.map {x =>x.toUpperCase}
upperCaseCities.collect.foreach(println)
TOKYO
NEW YORK
PARIS
SAN FRANCISCO
```

Let's show another example of map.

```
val lines = sc.parallelize(List("Michael Jordan", "iPhone"))

val words = lines.map(line =>line.split(" "))

words.collect

res2: Array[Array[String]] = Array(Array(Michael, Jordan), Array(iPhone))
```

FlatMap

FlatMap executes a function against each element in the RDD and then flattens the results.

```
val lines = sc.parallelize(List("Michael Jordan", "iPhone"))

val words = lines.flatMap(line =>line.split(" "))

words.collect

res3: Array[String] = Array(Michael, Jordan, iPhone)
```

Filter

Filter returns an RDD that only includes elements that match the condition specified.

```
val lines = sc.parallelize(List("Michael Jordan", "iPhone","Michael
Corleone"))

val words = lines.map(line =>line.split(" "))

val results = words.filter(w =>w.contains("Michael"))

results.collect

res9: Array[Array[String]] = Array(Array(Michael, Jordan), Array(Michael,
Corleone))
```

Distinct

Distinct returns only distinct values.

```
val cities1 = sc.parallelize(List("tokyo","tokyo","paris","sydney"))

val cities2 = sc.parallelize(List("perth","tokyo","canberra","sydney"))
```

```
val cities3 = cities1.union(cities2)

cities3.distinct.collect.foreach(println)

sydney
perth
canberra
tokyo
paris
```

ReduceByKey

ReduceByKey combines values with the same key using the specified reduce function.

```
val pairRDD = sc.parallelize(List(("a", 1), ("b",2), ("c",3), ("a", 30),
("b",25), ("a",20)))
val sumRDD = pairRDD.reduceByKey((x,y) =>x+y)
sumRDD.collect
res15: Array[(String, Int)] = Array((b,27), (a,51), (c,3))
```

Keys

Keys returns an RDD containing just the keys.

```
val rdd = sc.parallelize(List(("a", "Larry"), ("b", "Curly"), ("c", "Moe")))

val keys = rdd.keys

keys.collect.foreach(println)

a
b
c
```

Values

Values return an RDD containing just the values.

```
val rdd = sc.parallelize(List(("a", "Larry"), ("b", "Curly"), ("c", "Moe")))

val value = rdd.values
```

```
value.collect.foreach(println)

Larry
Curly
Moe
```

Inner Join

Inner join returns an RDD of all elements from both RDD based on the join predicate.

```
val data = Array((100,"Jim Hernandez"), (101,"Shane King"))
val employees = sc.parallelize(data)

val data2 = Array((100,"Glendale"), (101,"Burbank"))
val cities = sc.parallelize(data2)

val data3 = Array((100,"CA"), (101,"CA"), (102,"NY"))
val states = sc.parallelize(data3)

val record = employees.join(cities).join(states)

record.collect.foreach(println)

(100,((Jim Hernandez,Glendale),CA))
(101,((Shane King,Burbank),CA))
```

RightOuterJoin and LeftOuterJoin

RightOuterJoin returns an RDD of elements from the right RDD even if there are no matching rows on the left RDD. A LeftOuterJoin is equivalent to the RightOuterJoin with the columns in different order.

```
val record = employees.join(cities).rightOuterJoin(states)

record.collect.foreach(println)

(100,(Some((Jim Hernandez,Glendale)),CA))
(102,(None,NY))
(101,(Some((Shane King,Burbank)),CA))
```

Union

Union returns an RDD that contains the combination of two or more RDDs.

```
val  data = Array((103,"Mark Choi","Torrance","CA"), (104,"Janet
Reyes","RollingHills","CA"))
val employees = sc.parallelize(data)
val  data = Array((105,"Lester Cruz","VanNuys","CA"), (106,"John
White","Inglewood","CA"))
val employees2 = sc.parallelize(data)
val rdd = sc.union([employees, employees2])
rdd.collect.foreach(println)
(103,MarkChoi,Torrance,CA)
(104,JanetReyes,RollingHills,CA)
(105,LesterCruz,VanNuys,CA)
(106,JohnWhite,Inglewood,CA)
```

Subtract

Subtract returns an RDD that contains only the elements that are in the first RDD.

```
val data = Array((103,"Mark Choi","Torrance","CA"),  (104,"Janet
Reyes","Rolling Hills","CA"),(105,"Lester Cruz","Van Nuys","CA"))

val rdd = sc.parallelize(data)

val data2 = Array((103,"Mark Choi","Torrance","CA"))
val rdd2 = sc.parallelize(data2)

val employees = rdd.subtract(rdd2)

employees.collect.foreach(println)

(105,LesterCruz,Van Nuys,CA)
(104,JanetReyes,Rolling Hills,CA)
```

Coalesce

Coalesce reduces the number of partitions in an RDD. You might want to use coalesce after performing a filter on a large RDD. While filtering reduces the amount of data consumed by the new RDD, it inherits the number of partitions of the original RDD. If the new RDD is significantly smaller than the original RDD, it may have hundreds or thousands of small partitions, which could cause performance issues.

Coalesce is also useful when you want to reduce the number of files generated by Spark when writing to HDFS, preventing the dreaded "small file" problem. Each partition gets written as separate files to HDFS. Note that you might run into performance issues when using coalesce since you are effectively reducing the degree of parallelism while writing to HDFS. Try increasing the number of partitions if that happens. In the following example, we're writing only one Parquet file to HDFS.

```
df.coalesce(1).write.mode("append").parquet("/user/hive/warehouse/Mytable")
```

Repartition

Repartition can both decrease and increase the number of partitions in an RDD. You would generally use coalesce when reducing partitions since it's more efficient than repartition. Increasing the number of partitions is useful for increasing the degree of parallelism when writing to HDFS. In the following example, we're writing six Parquet files to HDFS.

```
df.repartition(6).write.mode("append").parquet("/user/hive/warehouse/
Mytable")
```

Note Coalesce is generally faster than repartition. Repartition will perform a full shuffle, creating new partitions and equally distributing data across worker nodes. Coalesce minimizes data movement and avoids a full shuffle by using existing partitions.

Actions

An action is an RDD operation that returns a value to the driver program. I list some of the most common actions. Refer to the online Spark documentation for a complete list of actions.

Collect

Collect returns the entire dataset as an array to the driver program.

```
val myCities = sc.parallelize(List("tokyo","new york","paris","san
francisco"))
myCities.collect
res2: Array[String] = Array(tokyo, new york, paris, san francisco)
```

Count

Count returns a count of the number of elements in the dataset.

```
val myCities = sc.parallelize(List("tokyo","new york","paris","san
francisco"))
myCities.count
res3: Long = 4
```

Take

Take returns the first *n* elements of the dataset as an array.

```
val myCities = sc.parallelize(List("tokyo","new york","paris","san
francisco"))
myCities.take(2)
res4: Array[String] = Array(tokyo, new york)
```

Foreach

Foreach executes a function on each element of the dataset.

```
val myCities = sc.parallelize(List("tokyo","new york","paris","san
francisco"))

myCities.collect.foreach(println)

tokyo
newyork
paris
sanFrancisco
```

Lazy Evaluation

Spark supports lazy evaluation, which is critical for big data processing. All transformations in Spark are lazily evaluated. Spark does not execute transformations immediately. You can continue to define more transformations. When you finally want the final results, you execute an action, which causes the transformations to be executed.

Caching

Each transformation is re-executed each time you run an action by default. You can cache an RDD in memory using the cache or persist method to avoid re-executing the transformation multiple times.

Accumulator

Accumulators are variables that are only "added" to. They are usually used to implement counters. In the example, I add up the elements of an array using an accumulator:

```
val accum = sc.longAccumulator("Accumulator 01")

sc.parallelize(Array(10, 20, 30, 40)).foreach(x =>accum.add(x))

accum.value
res2: Long = 100
```

Broadcast Variable

Broadcast variables are read-only variables stored on each node's memory. Spark uses high-speed broadcast algorithms to reduce network latency of copying broadcast variables. Instead of storing data in a slow storage engine such as HDFS or S3, using broadcast variables is a faster way to store a copy of a dataset on each node.

```
val broadcastVar = sc.broadcast(Array(10, 20, 30))

broadcastVar.value
res0: Array[Int] = Array(10, 20, 30)
```

Spark SQL, Dataset, and DataFrames API

Spark SQL was developed to make it easier to process and analyze structured data. Dataset is similar to an RDD in that it supports strong typing, but under the hood Dataset has a much more efficient engine. Starting in Spark 2.0, the Dataset API is now the primary programming interface. The DataFrame is just a Dataset with named columns, similar to relational table. Together, Spark SQL and DataFrames provide a powerful programming interface for processing and analyzing structured data. Here's a quick example on how to use the DataFrames API.

```
val jsonDF = spark.read.json("/jsondata/customers.json")

jsonDF.show
+---+------+--------------+-----+------+-----+
|age|  city|          name|state|userid|  zip|
+---+------+--------------+-----+------+-----+
| 35|Frisco| Jonathan West|   TX|   200|75034|
| 28|Dallas|Andrea Foreman|   TX|   201|75001|
| 69| Plano|  Kirsten Jung|   TX|   202|75025|
| 52| Allen|Jessica Nguyen|   TX|   203|75002|
+---+------+--------------+-----+------+-----+

jsonDF.select ("age","city").show

+---+------+
|age|  city|
+---+------+
| 35|Frisco|
| 28|Dallas|
| 69| Plano|
| 52| Allen|
+---+------+
```

```
jsonDF.filter($"userid" < 202).show()

+---+------+--------------+-----+------+-----+
|age|  city|          name|state|userid|  zip|
+---+------+--------------+-----+------+-----+
| 35|Frisco| Jonathan West|   TX|   200|75034|
| 28|Dallas|Andrea Foreman|   TX|   201|75001|
+---+------+--------------+-----+------+-----+

jsonDF.createOrReplaceTempView("jsonDF")

val df = spark.sql("SELECT userid, zip FROM jsonDF")

df.show
+------+-----+
|userid|  zip|
+------+-----+
|   200|75034|
|   201|75001|
|   202|75025|
|   203|75002|
+------+-----+
```

Note The DataFrame and Dataset APIs have been unified in Spark 2.0. The DataFrame is now just a type alias for a Dataset of Row, where a Row is a generic untyped object. In contrast, Dataset is a collection of strongly typed objects Dataset[T]. Scala supports strongly typed and untyped API, while in Java, Dataset[T] is the main abstraction. DataFrames is the main programming interface for R and Python due to its lack of support for compile-time type safety.

Spark Data Sources

Reading and writing to different file formats and data sources is one of the most common data processing tasks. We will use both the RDD and DataFrames API in our examples.

CSV

Spark provides you with different ways to read data from CSV files. You can read the data into an RDD first and then convert it to DataFrame.

```
val dataRDD = sc.textFile("/sparkdata/customerdata.csv")
val parsedRDD = dataRDD.map{_.split(",")}
case class CustomerData(customerid: Int, name: String, city: String, state:
String, zip: String)
val dataDF = parsedRDD.map{ a =>CustomerData (a(0).toInt, a(1).toString,
a(2).toString,a(3).toString,a(4).toString) }.toDF
```

Starting in Spark 2.0, the CSV connector is already built-in.

```
val dataDF = spark.read.format("csv")
             .option("header", "true")
             .load("/sparkdata/customerdata.csv")
```

XML

Databricks has a Spark XML package that makes it easy to read XML data.

```
cat users.xml
```

```
<userid>100</userid><name>Wendell Ryan</name><city>San Diego</city>
<state>CA</state><zip>92102</zip>
<userid>101</userid><name>Alicia Thompson</name><city>Berkeley</city>
<state>CA</state><zip>94705</zip>
<userid>102</userid><name>Felipe Drummond</name><city>Palo Alto</city>
<state>CA</state><zip>94301</zip>
<userid>103</userid><name>Teresa Levine</name><city>Walnut Creek</city>
<state>CA</state><zip>94507</zip>
```

```
hadoop fs -mkdir /xmldata
hadoop fs -put users.xml /xmldata
```

```
spark-shell --packages  com.databricks:spark-xml_2.10:0.4.1
```

Create a DataFrame using Spark XML. In this example, we specify the row tag and the path in HDFS where the XML file is located.

```
import com.databricks.spark.xml._

val xmlDF = spark.read
            .option("rowTag", "user")
            .xml("/xmldata/users.xml");

xmlDF: org.apache.spark.sql.DataFrame = [city: string, name: string,
state: string, userid: bigint, zip: bigint]
```

Let's also take a look at the data.

```
xmlDF.show

+------------+---------------+-----+------+-----+
|        city|           name|state|userid|  zip|
+------------+---------------+-----+------+-----+
|   San Diego|   Wendell Ryan|   CA|   100|92102|
|    Berkeley|Alicia Thompson|   CA|   101|94705|
|   Palo Alto|Felipe Drummond|   CA|   102|94301|
|Walnut Creek|  Teresa Levine|   CA|   103|94507|
+------------+---------------+-----+------+-----+
```

JSON

We'll create a JSON file as a sample data for this example. Make sure the file is in a folder in HDFS called /jsondata.

```
cat users.json

{"userid": 200, "name": "Jonathan West", "city":"Frisco", "state":"TX",
"zip": "75034", "age":35}
{"userid": 201, "name": "Andrea Foreman", "city":"Dallas", "state":"TX",
"zip": "75001", "age":28}
{"userid": 202, "name": "Kirsten Jung", "city":"Plano", "state":"TX",
"zip": "75025", "age":69}
{"userid": 203, "name": "Jessica Nguyen", "city":"Allen", "state":"TX",
"zip": "75002", "age":52}
```

Create a DataFrame from the JSON file.

```
val jsonDF = spark.read.json("/jsondata/users.json")

jsonDF: org.apache.spark.sql.DataFrame = [age: bigint, city: string,
name: string, state: string, userid: bigint, zip: string]
```

Check the data.

```
jsonDF.show

+---+------+-------------+-----+------+-----+
|age|  city|         name|state|userid|  zip|
+---+------+-------------+-----+------+-----+
| 35|Frisco| Jonathan West|   TX|   200|75034|
| 28|Dallas|Andrea Foreman|   TX|   201|75001|
| 69| Plano|  Kirsten Jung|   TX|   202|75025|
| 52| Allen|Jessica Nguyen|   TX|   203|75002|
+---+------+-------------+-----+------+-----+
```

Relational and MPP Databases

We use MySQL in this example, but other relational databases and MPP engines such as Oracle, Snowflake, Redshift, Impala, Presto, and Azure DW are also supported. Generally, as long as the relational database has a JDBC driver, it should be accessible from Spark. Performance is dependent on your JDBC driver's support for batch operations. Please check your JDBC driver's documentation for more details.

```
mysql -u root -pmypassword

create databases salesdb;

use salesdb;

create table customers (
customerid INT,
name VARCHAR(100),
city VARCHAR(100),
state CHAR(3),
zip  CHAR(5));

spark-shell --driver-class-path mysql-connector-java-5.1.40-bin.jar
```

Start the spark-shell.

Read the CSV file into an RDD and convert it to a DataFrame.

```
val dataRDD = sc.textFile("/home/hadoop/test.csv")
val parsedRDD = dataRDD.map{_.split(",")}

case class CustomerData(customerid: Int, name: String, city: String, state:
String, zip: String)

val dataDF = parsedRDD.map{ a =>CustomerData (a(0).toInt, a(1).toString,
a(2).toString,a(3).toString,a(4).toString) }.toDF
```

Register the data frame as a temp table so that we can run SQL queries against it.

```
dataDF.createOrReplaceTempView("dataDF")
```

Let's set up the connection properties.

```
val jdbcUsername = "myuser"
val jdbcPassword = "mypass"
val jdbcHostname = "10.0.1.112"
val jdbcPort = 3306
val jdbcDatabase ="salesdb"
val jdbcrewriteBatchedStatements = "true"
val jdbcUrl = s"jdbc:mysql://${jdbcHostname}:${jdbcPort}/${jdbcDatabase}?us
er=${jdbcUsername}&password=${jdbcPassword}&rewriteBatchedStatements=${jdbc
rewriteBatchedStatements}"

val connectionProperties = new java.util.Properties()
```

This will allow us to specify the correct save mode - append, overwrite, and so on.

```
import org.apache.spark.sql.SaveMode
```

Insert the data returned by the SELECT statement to the customer table stored in the MySQL salesdb database.

```
spark.sql("select * from dataDF")
         .write
         .mode(SaveMode.Append)
         .jdbc(jdbcUrl, "customers", connectionProperties)
```

Let's read a table using JDBC. Let's populate the users table in MySQL with some test data. Make sure the users table exists in the salesdb database.

```
mysql -u root -pmypassword

use salesdb;

describe users;
+--------+--------------+------+-----+---------+-------+
| Field  | Type         | Null | Key | Default | Extra |
+--------+--------------+------+-----+---------+-------+
| userid | bigint(20)   | YES  |     | NULL    |       |
| name   | varchar(100) | YES  |     | NULL    |       |
| city   | varchar(100) | YES  |     | NULL    |       |
| state  | char(3)      | YES  |     | NULL    |       |
| zip    | char(5)      | YES  |     | NULL    |       |
| age    | tinyint(4)   | YES  |     | NULL    |       |
+--------+--------------+------+-----+---------+-------+

select * from users;
Empty set (0.00 sec)

insert into users values (300,'Fred Stevens','Torrance','CA',90503,23);

insert into users values (301,'Nancy Gibbs','Valencia','CA',91354,49);

insert into users values (302,'Randy Park','Manhattan Beach','CA',90267,21);

insert into users values (303,'Victoria Loma','Rolling Hills','CA',90274,75);

select * from users;
+--------+---------------+-----------------+-------+-------+------+
| userid | name          | city            | state | zip   | age  |
+--------+---------------+-----------------+-------+-------+------+
|    300 | Fred Stevens  | Torrance        | CA    | 90503 |   23 |
|    301 | Nancy Gibbs   | Valencia        | CA    | 91354 |   49 |
|    302 | Randy Park    | Manhattan Beach | CA    | 90267 |   21 |
|    303 | Victoria Loma | Rolling Hills   | CA    | 90274 |   75 |
+--------+---------------+-----------------+-------+-------+------+

spark-shell --driver-class-path mysql-connector-java-5.1.40-bin.jar --jars
mysql-connector-java-5.1.40-bin.jar
```

Let's set up the jdbcurl and connection properties.

```
val jdbcURL = s"jdbc:mysql://10.0.1.101:3306/salesdb?user=myuser&password=
mypass"
```

```
val connectionProperties = new java.util.Properties()
```

We can create a DataFrame from an entire table.

```
val df = spark.read.jdbc(jdbcURL, "users", connectionProperties)
```

```
df.show
+------+------------+--------------+-----+-----+---+
|userid|        name|          city|state|  zip|age|
+------+------------+--------------+-----+-----+---+
|   300| Fred Stevens|      Torrance|   CA|90503| 23|
|   301|  Nancy Gibbs|      Valencia|   CA|91354| 49|
|   302|   Randy Park|Manhattan Beach|   CA|90267| 21|
|   303|Victoria Loma| Rolling Hills|   CA|90274| 75|
+------+------------+--------------+-----+-----+---+
```

Parquet

Reading and writing to Parquet is straightforward.

```
val df = spark.read.load("/sparkdata/employees.parquet")
```

```
df.select("id","firstname","lastname","salary")
          .write
          .format("parquet")
          .save("/sparkdata/myData.parquet")
```

```
You can run SELECT statements on Parquet files directly.
```

```
val df = spark.sql("SELECT * FROM parquet.`/sparkdata/myData.parquet`")
```

HBase

There are multiple ways to access HBase from Spark. For example, you can use SaveAsHadoopDataset to write data to HBase. Start the HBase shell.

Create an HBase table and populate it with test data.

```
hbase shell

create 'users', 'cf1'
```

Start the spark-shell.

```
spark-shell

val hconf = HBaseConfiguration.create()
val jobConf = new JobConf(hconf, this.getClass)
jobConf.setOutputFormat(classOf[TableOutputFormat])
jobConf.set(TableOutputFormat.OUTPUT_TABLE,"users")

val num = sc.parallelize(List(1,2,3,4,5,6))

val theRDD = num.filter.map(x=>{

      val rowkey = "row" + x

val put = new Put(Bytes.toBytes(rowkey))

      put.add(Bytes.toBytes("cf1"), Bytes.toBytes("fname"), Bytes.
      toBytes("my fname" + x))

    (newImmutableBytesWritable, put)
})
theRDD.saveAsHadoopDataset(jobConf)
```

You can also use the HBase client API from Spark to read and write data to HBase. As discussed earlier, Scala has access to all Java libraries.

Start the HBase shell. Create another HBase table and populate it with test data.

```
hbase shell

create 'employees', 'cf1'

put 'employees','400','cf1:name', 'Patrick Montalban'
put 'employees','400','cf1:city', 'Los Angeles'
put 'employees','400','cf1:state', 'CA'
put 'employees','400','cf1:zip', '90010'
put 'employees','400','cf1:age', '71'
```

```
put 'employees','401','cf1:name', 'Jillian Collins'
put 'employees','401','cf1:city', 'Santa Monica'
put 'employees','401','cf1:state', 'CA'
put 'employees','401','cf1:zip', '90402'
put 'employees','401','cf1:age', '45'

put 'employees','402','cf1:name', 'Robert Sarkisian'
put 'employees','402','cf1:city', 'Glendale'
put 'employees','402','cf1:state', 'CA'
put 'employees','402','cf1:zip', '91204'
put 'employees','402','cf1:age', '29'

put 'employees','403','cf1:name', 'Warren Porcaro'
put 'employees','403','cf1:city', 'Burbank'
put 'employees','403','cf1:state', 'CA'
put 'employees','403','cf1:zip', '91523'
put 'employees','403','cf1:age', '62'
```

Let's verify if the data was successfully inserted into our HBase table.

```
scan 'employees'

ROW         COLUMN+CELL
 400        column=cf1:age, timestamp=1493105325812, value=71
 400        column=cf1:city, timestamp=1493105325691, value=Los Angeles
 400        column=cf1:name, timestamp=1493105325644, value=Patrick Montalban
 400        column=cf1:state, timestamp=1493105325738, value=CA
 400        column=cf1:zip, timestamp=1493105325789, value=90010
 401        column=cf1:age, timestamp=1493105334417, value=45
 401        column=cf1:city, timestamp=1493105333126, value=Santa Monica
 401        column=cf1:name, timestamp=1493105333050, value=Jillian Collins
 401        column=cf1:state, timestamp=1493105333145, value=CA
 401        column=cf1:zip, timestamp=1493105333165, value=90402
 402        column=cf1:age, timestamp=1493105346254, value=29
 402        column=cf1:city, timestamp=1493105345053, value=Glendale
 402        column=cf1:name, timestamp=1493105344979, value=Robert Sarkisian
 402        column=cf1:state, timestamp=1493105345074, value=CA
```

```
402       column=cf1:zip, timestamp=1493105345093, value=91204
403       column=cf1:age, timestamp=1493105353650, value=62
403       column=cf1:city, timestamp=1493105352467, value=Burbank
403       column=cf1:name, timestamp=1493105352445, value=Warren Porcaro
403       column=cf1:state, timestamp=1493105352513, value=CA
403       column=cf1:zip, timestamp=1493105352549, value=91523
```

Start the spark-shell.

```
spark-shell

import org.apache.hadoop.fs.Path;
import org.apache.hadoop.hbase.{HBaseConfiguration, HTableDescriptor}
import org.apache.hadoop.hbase.client.HBaseAdmin
import org.apache.hadoop.hbase.mapreduce.TableInputFormat
import org.apache.hadoop.hbase.HColumnDescriptor
import org.apache.hadoop.hbase.client.Put;
import org.apache.hadoop.hbase.client.Get;
import org.apache.hadoop.hbase.client.HTable;
import org.apache.hadoop.conf.Configuration;
import org.apache.hadoop.hbase.client.Result;
import org.apache.hadoop.hbase.util.Bytes;
import java.io.IOException;

val configuration = HBaseConfiguration.create()
```

Specify the HBase table and rowkey.

```
val table = new HTable(configuration, "employees");
val g = new Get(Bytes.toBytes("401"))
val result = table.get(g);
```

Extract the values from the table.

```
val val2 = result.getValue(Bytes.toBytes("cf1"),Bytes.toBytes("name"));
val val3 = result.getValue(Bytes.toBytes("cf1"),Bytes.toBytes("city"));
val val4 = result.getValue(Bytes.toBytes("cf1"),Bytes.toBytes("state"));
val val5 = result.getValue(Bytes.toBytes("cf1"),Bytes.toBytes("zip"));
val val6 = result.getValue(Bytes.toBytes("cf1"),Bytes.toBytes("age"));
```

Convert the values to the appropriate data types.

```
val id = Bytes.toString(result.getRow())
val name = Bytes.toString(val2);
val city = Bytes.toString(val3);
val state = Bytes.toString(val4);
val zip = Bytes.toString(val5);
val age = Bytes.toShort(val6);
```

Print the values.

```
println(" employee id: " + id + " name: " + name + " city: " + city + "
state: " + state + " zip: " + zip + " age: " + age);
```

```
employee id: 401 name: Jillian Collins city: Santa Monica state: CA zip:
90402 age: 13365
```

Let's write to HBase using the HBase API.

```
val configuration = HBaseConfiguration.create()
val table = new HTable(configuration, "employees");
```

Specify a new rowkey.

```
val p = new Put(new String("404").getBytes());
```

Populate the cells with the new values.

```
p.add("cf1".getBytes(), "name".getBytes(), new String("Denise Shulman").
getBytes());
p.add("cf1".getBytes(), "city".getBytes(), new String("La Jolla").
getBytes());
p.add("cf1".getBytes(), "state".getBytes(), new String("CA").getBytes());
p.add("cf1".getBytes(), "zip".getBytes(), new String("92093").getBytes());
p.add("cf1".getBytes(), "age".getBytes(), new String("56").getBytes());
```

Write to the HBase table.

```
table.put(p);
table.close();
```

Confirm that the values were successfully inserted into the HBase table.

Start the HBase shell.

```
hbase shell

scan 'employees'

ROW       COLUMN+CELL
 400      column=cf1:age, timestamp=1493105325812, value=71
 400      column=cf1:city, timestamp=1493105325691, value=Los Angeles
 400      column=cf1:name, timestamp=1493105325644, value=Patrick Montalban
 400      column=cf1:state, timestamp=1493105325738, value=CA
 400      column=cf1:zip, timestamp=1493105325789, value=90010
 401      column=cf1:age, timestamp=1493105334417, value=45
 401      column=cf1:city, timestamp=1493105333126, value=Santa Monica
 401      column=cf1:name, timestamp=1493105333050, value=Jillian Collins
 401      column=cf1:state, timestamp=1493105333145, value=CA
 401      column=cf1:zip, timestamp=1493105333165, value=90402
 402      column=cf1:age, timestamp=1493105346254, value=29
 402      column=cf1:city, timestamp=1493105345053, value=Glendale
 402      column=cf1:name, timestamp=1493105344979, value=Robert Sarkisian
 402      column=cf1:state, timestamp=1493105345074, value=CA
 402      column=cf1:zip, timestamp=1493105345093, value=91204
 403      column=cf1:age, timestamp=1493105353650, value=62
 403      column=cf1:city, timestamp=1493105352467, value=Burbank
 403      column=cf1:name, timestamp=1493105352445, value=Warren Porcaro
 403      column=cf1:state, timestamp=1493105352513, value=CA
 403      column=cf1:zip, timestamp=1493105352549, value=91523
 404      column=cf1:age, timestamp=1493123890714, value=56
 404      column=cf1:city, timestamp=1493123890714, value=La Jolla
 404      column=cf1:name, timestamp=1493123890714, value=Denise Shulman
 404      column=cf1:state, timestamp=1493123890714, value=CA
 404      column=cf1:zip, timestamp=1493123890714, value=92093
```

Although generally slower, you can also access HBase via a SQL query engine such as Impala or Presto.

Amazon S3

Amazon S3 is a popular object store frequently used as data store for transient clusters. It's also a cost-effective storage for backups and cold data. Reading data from S3 is just like reading data from HDFS or any other file system.

Read a CSV file from Amazon S3. Make sure you've configured your S3 credentials.

```
val myCSV = sc.textFile("s3a://mydata/customers.csv")
```

Map CSV data to an RDD.

```
import org.apache.spark.sql.Row

val myRDD = myCSV.map(_.split(',')).map(e ⇒ Row(r(0).trim.toInt, r(1),
r(2).trim.toInt, r(3)))
```

Create a schema.

```
import org.apache.spark.sql.types.{StructType, StructField, StringType,
IntegerType};

val mySchema = StructType(Array(
StructField("customerid",IntegerType,false),
StructField("customername",StringType,false),
StructField("age",IntegerType,false),
StructField("city",StringType,false)))

val myDF = spark.createDataFrame(myRDD, mySchema)
```

Solr

You can interact with Solr from Spark using SolrJ.[viii]

```
import java.net.MalformedURLException;
import org.apache.solr.client.solrj.SolrServerException;
import org.apache.solr.client.solrj.impl.HttpSolrServer;
import org.apache.solr.client.solrj.SolrQuery;
import org.apache.solr.client.solrj.response.QueryResponse;
import org.apache.solr.common.SolrDocumentList;
```

```
val solr = new HttpSolrServer("http://master02:8983/solr/mycollection");

val query = new SolrQuery();

query.setQuery("*:*");
query.addFilterQuery("userid:3");
query.setFields("userid","name","age","city");
query.setStart(0);
query.set("defType", "edismax");

val response = solr.query(query);
val results = response.getResults();

println(results);
```

A much easier way to access Solr collections from Spark is through the spark-solr package. Lucidworks started the spark-solr project to provide Spark-Solr integration.[ix] Using spark-solr is so much easier and powerful compared to SolrJ, allowing you to create DataFrames from Solr collections.

Start by importing the JAR file from spark-shell.

```
spark-shell --jars spark-solr-3.0.1-shaded.jar
```

Specify the collection and connection information.

```
val options = Map( "collection" -> "mycollection","zkhost" -> "{
master02:8983/solr}")
```

Create a DataFrame.

```
val solrDF = spark.read.format("solr")
      .options(options)
      .load
```

Microsoft Excel

While accessing Excel spreadsheets from Spark is something that I would not normally recommend, certain use cases require the capability. A company called Crealytics developed a Spark plug-in for interacting with Excel. The library requires Spark 2.x. The package can be added using the --packages command-line option.

```
spark-shell --packages com.crealytics:spark-excel_2.11:0.9.12
```

Create a DataFrame from an Excel worksheet.

```
val ExcelDF = spark.read
    .format("com.crealytics.spark.excel")
    .option("sheetName", "sheet1")
    .option("useHeader", "true")
    .option("inferSchema", "true")
    .option("treatEmptyValuesAsNulls", "true")
    .load("budget.xlsx")
```

Write a DataFrame to an Excel worksheet.

```
ExcelDF2.write
  .format("com.crealytics.spark.excel")
  .option("sheetName", "sheet1")
  .option("useHeader", "true")
  .mode("overwrite")
  .save("budget2.xlsx")
```

You can find more details from their GitHub page: github.com/crealytics.

Secure FTP

Downloading files from and writing DataFrames to an SFTP server is also a popular request. SpringML provides a Spark SFTP connector library. The library requires Spark 2.x and utilizes jsch, a Java implementation of SSH2. Reading from and writing to SFTP servers will be executed as a single process.

```
spark-shell --packages com.springml:spark-sftp_2.11:1.1.
```

Create a DataFrame from the file in SFTP server.

```
val sftpDF = spark.read.
           format("com.springml.spark.sftp").
           option("host", "sftpserver.com").
           option("username", "myusername").
           option("password", "mypassword").
           option("inferSchema", "true").
           option("fileType", "csv").
```

```
        option("delimiter", ",").
        load("/myftp/myfile.csv")
```

Write DataFrame as CSV file to FTP server.

```
sftpDF2.write.
    format("com.springml.spark.sftp").
    option("host", "sftpserver.com").
    option("username", "myusername").
    option("password", "mypassword").
    option("fileType", "csv").
    option("delimiter", ",").
    save("/myftp/myfile.csv")
```

You can find more details from their GitHub page: `github.com/springml/spark-sftp`.

Introduction to Spark MLlib

Machine learning is one of Spark's main applications. Spark MLlib includes popular machine learning algorithms for regression, classification, clustering, collaborative filtering, and frequent pattern mining. It also provides a wide set of features for building pipelines, model selection and tuning, and feature selection, extraction, and transformation.

Spark MLlib Algorithms

Spark MLlib includes a plethora of machine learning algorithms for various tasks. We cover most of them in succeeding chapters.

Classification

- Logistic Regression (Binomial and Multinomial)
- Decision Tree
- Random Forest
- Gradient-Boosted Tree
- Multilayer Perceptron

- Linear Support Vector Machine
- Naïve Bayes
- One-vs.-Rest

Regression

- Linear Regression
- Decision Tree
- Random Forest
- Gradient-Boosted Tree
- Survival Regression
- Isotonic Regression

Clustering

- K-Means
- Bisecting K-Means
- Gaussian Mixture Model
- Latent Dirichlet Allocation (LDA)

Collaborative Filtering

- Alternating Least Square (ALS)

Frequent Pattern Mining

- FP-Growth
- PrefixSpan

ML Pipelines

Early versions of Spark MLlib only included an RDD-based API. The DataFrame-based API is now the primary API for Spark. The RDD-based API will be deprecated in Spark 2.3 once the DataFrames-based API reaches feature parity.[x] The RDD-based API will be removed in Spark 3.0. The DataFrames-based API makes it easy to transform features by providing a higher-level abstraction for representing tabular data similar to a relational database table, making it a natural choice for implementing pipelines.

The Spark MLlib API introduces several concepts for creating machine learning pipelines. Figure 2-3 shows a simple Spark MLlib pipeline for processing text data. A tokenizer breaks the text into a bag of word, appending the words to the output DataFrame. Term frequency–inverse document frequency (TF–IDF) takes the DataFrame as input, converts the bag of words into a feature vector, and adds them to the third DataFrame.

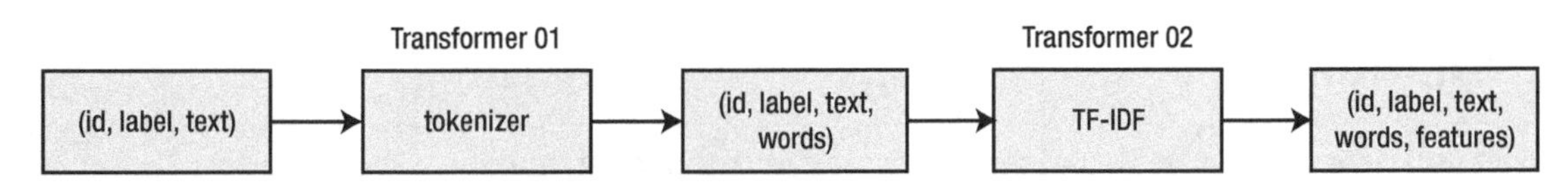

Figure 2-3. *A simple Spark MLlib pipeline*

Pipeline

A pipeline is a sequence of connected stages to create a machine learning workflow. A stage can be either a transformer or estimator.

Transformer

A transformer takes a DataFrame as input and outputs a new DataFrame with additional columns appended to the new DataFrame. The new DataFrame includes the columns from the input DataFrame and the additional columns.

Estimator

An estimator is a machine learning algorithm that fits a model on training data. Estimators accept training data and produce a machine learning model.

ParamGridBuilder

A ParamGridBuilder is used to build a parameter grid. The CrossValidator performs a grid search and trains models with a combination of user-specified hyperparameters in the parameter grid.

CrossValidator

A CrossValidator cross-evaluates fitted machine learning models and outputs the best one by trying to fit the underlying estimator with user-specified combinations of hyperparameters. Model selection is performed with the CrossValidator or TrainValidationSplit estimators.

Evaluator

An evaluator calculates the performance of your machine learning models. It outputs a metric such as precision and recall to measure how well a fitted model performs. Examples of evaluators include BinaryClassificationEvaluator and MulticlassClassificationEvaluator for binary and multiclass classification tasks, respectively, and RegressionEvaluator for regression tasks.

Feature Extraction, Transformation, and Selection

Most of the time additional preprocessing is needed before raw data can be used to fit models. Distance-based algorithms, for instance, require features to be standardized. Some algorithms perform better when categorical data are one-hot encoded. Text data usually require tokenization and feature vectorization. Dimensionality reduction might be needed for extremely large datasets. Spark MLlib includes an extensive collection of transformers and estimators for these types of tasks. I'll discuss some of the most commonly used transformers and estimators available in Spark MLlib.

StringIndexer

Most machine learning algorithms cannot work with strings directly and require data to be in numeric format. StringIndexer is an estimator that converts string columns of labels to indices. It supports four different ways to generate the indices: alphabetDesc, alphabetAsc, frequencyDesc, and frequencyAsc. The default is set to frequencyDesc, with the most frequent label set to 0, and the results are sorted by descending order of label frequency.

```
import org.apache.spark.ml.feature.StringIndexer

val df = spark.createDataFrame(
  Seq((0, "car"), (1, "car"), (2, "truck"), (3, "van"), (4, "van"),
  (5, "van"))
).toDF("id", "class")

df.show
+---+-----+
| id|class|
+---+-----+
|  0|  car|
|  1|  car|
|  2|truck|
|  3|  van|
|  4|  van|
|  5|  van|
+---+-----+
val model = new StringIndexer()
            .setInputCol("class")
            .setOutputCol("classIndex")

val indexer = model.fit(df)

val indexed = indexer.transform(df)

indexed.show()
+---+-----+----------+
| id|class|classIndex|
+---+-----+----------+
|  0|  car|       1.0|
|  1|  car|       1.0|
|  2|truck|       2.0|
|  3|  van|       0.0|
|  4|  van|       0.0|
|  5|  van|       0.0|
+---+-----+----------+
```

Tokenizer

When analyzing text data, it is usually essential to split sentences into individual terms or words. A tokenizer does exactly that. You can perform more advanced tokenization using regular expression using RegexTokenizer. Tokenization is usually one of the first steps in a machine learning NLP pipeline. I will discuss natural language processing (NLP) in greater detail in Chapter 4.

```
import org.apache.spark.ml.feature.Tokenizer

val df = spark.createDataFrame(Seq(
  (0, "Mark gave a speech last night in Laguna Beach"),
  (1, "Oranges are full of nutrients and low in calories"),
  (2, "Eddie Van Halen is amazing")
)).toDF("id", "sentence")

 df.show(false)

+---+-------------------------------------------------+
|id |sentence                                         |
+---+-------------------------------------------------+
|0  |Mark gave a speech last night in Laguna Beach    |
|1  |Oranges are full of nutrients and low in calories|
|2  |Eddie Van Halen is amazing                       |
+---+-------------------------------------------------+

val tokenizer = new Tokenizer().setInputCol("sentence").
setOutputCol("words")

val tokenized = tokenizer.transform(df)

tokenized.show(false)

+---+-------------------------------------------------+
|id |sentence                                         |
+---+-------------------------------------------------+
|0  |Mark gave a speech last night in Laguna Beach    |
|1  |Oranges are full of nutrients and low in calories|
|2  |Eddie Van Halen is amazing                       |
+---+-------------------------------------------------+
```

```
+------------------------------------------------------------+
|words                                                       |
+------------------------------------------------------------+
|[mark, gave, a, speech, last, night, in, laguna, beach]     |
|[oranges, are, full, of, nutrients, and, low, in, calories]|
|[eddie, van, halen, is, amazing]                            |
+------------------------------------------------------------+
```

VectorAssembler

Spark MLlib algorithms require that features be stored in a single vector column. Usually, training data will come in tabular format where data is stored in separate columns. VectorAssembler is a transformer that merges a set of columns into a single vector column.

```
import org.apache.spark.ml.feature.VectorAssembler

val df = spark.createDataFrame(
  Seq((0, 50000, 7, 1))
).toDF("id", "income", "employment_length", "marital_status")

val assembler = new VectorAssembler()
.setInputCols(Array("income", "employment_length", "marital_status"))
.setOutputCol("features")

val df2 = assembler.transform(df)

df2.show(false)

+---+------+-----------------+--------------+-----------------+
|id |income|employment_length|marital_status|features         |
+---+------+-----------------+--------------+-----------------+
|0  |50000 |7                |1             |[50000.0,7.0,1.0]|
+---+------+-----------------+--------------+-----------------+
```

StandardScaler

As discussed in Chapter 1, some machine learning algorithms require features to be normalized to work properly. StandardScaler is an estimator that normalizes features to have unit standard deviation and/or zero mean. It accepts two parameters: *withStd and withMean. withStd* scales the features to unit standard deviation. This parameter is set to true by default. Setting *withMean* to true centers the data with mean prior to scaling. This parameter is set to false by default.

```
import org.apache.spark.ml.feature.StandardScaler
import org.apache.spark.ml.feature.VectorAssembler

val df = spark.createDataFrame(
  Seq((0, 186, 200, 56),(1, 170, 198, 42))
).toDF("id", "height", "weight", "age")

val assembler = new VectorAssembler()
.setInputCols(Array("height", "weight", "age"))
.setOutputCol("features")

val df2 = assembler.transform(df)

df2.show(false)

+---+------+------+---+------------------+
|id |height|weight|age|features          |
+---+------+------+---+------------------+
|0  |186   |200   |56 |[186.0,200.0,56.0]|
|1  |170   |198   |42 |[170.0,198.0,42.0]|
+---+------+------+---+------------------+

val scaler = new StandardScaler()
  .setInputCol("features")
  .setOutputCol("scaledFeatures")
  .setWithStd(true)
  .setWithMean(false)

val model = scaler.fit(df2)

val scaledData = model.transform(df2)
```

```
scaledData.select("features","scaledFeatures").show(false)
+-----------------+---------------------------------------------------------+
|features         |scaledFeatures                                           |
+-----------------+---------------------------------------------------------+
|[186.0,200.0,56.0]|[16.440232662587228,141.42135623730948,5.656854249492]|
|[170.0,198.0,42.0]|[15.026019100214134,140.0071426749364,4.2426406871192]|
+-----------------+---------------------------------------------------------+
```

Additional transformers for rescaling data include Normalizer, MinMaxScaler, and MaxAbsScaler. Please check the Apache Spark online documentation for more details.

StopWordsRemover

Often used in text analysis, StopWordsRemover removes stop words from a sequence of strings. Stop words, such as I, the, and a, do not contribute a lot to the meaning of a document.

```
import org.apache.spark.ml.feature.StopWordsRemover

val remover = new StopWordsRemover().setInputCol("data").
setOutputCol("output")

val dataSet = spark.createDataFrame(Seq(
  (0, Seq("She", "is", "a", "cute", "baby")),
  (1, Seq("Bob", "never", "went", "to", "Seattle"))
)).toDF("id", "data")

val df = remover.transform(dataSet)

df.show(false)

+---+-------------------------------+--------------------------+
|id |data                           |output                    |
+---+-------------------------------+--------------------------+
|0  |[She, is, a, cute, baby]       |[cute, baby]              |
|1  |[Bob, never, went, to, Seattle]|[Bob, never, went, Seattle]|
+---+-------------------------------+--------------------------+
```

n-gram

When performing text analysis, it is sometimes advantageous to combine terms into n-grams, a combination of terms in a document. Creating n-grams helps extract more meaningful information from a document. For example, the words "San" and "Diego" individually don't mean much, but combining them into a bigram "San Diego" provides more context. We use n-gram later in Chapter 4.

```
import org.apache.spark.ml.feature.NGram

val df = spark.createDataFrame(Seq(
  (0, Array("Los", "Angeles", "Lobos", "San", "Francisco")),
  (1, Array("Stand", "Book", "Case", "Phone", "Mobile", "Magazine")),
  (2, Array("Deep", "Learning", "Machine", "Algorithm", "Pizza"))
)).toDF("id", "words")

val ngram = new NGram().setN(2).setInputCol("words").setOutputCol("ngrams")

val df2 = ngram.transform(df)

df2.select("ngrams").show(false)

+------------------------------------------------------------------+
|ngrams                                                            |
+------------------------------------------------------------------+
|[Los Angeles, Angeles Lobos, Lobos San, San Francisco]            |
|[Stand Book, Book Case, Case Phone, Phone Mobile, Mobile Magazine]  |
|[Deep Learning, Learning Machine, Machine Algorithm, Algorithm Pizza]|
+------------------------------------------------------------------+
```

OneHotEncoderEstimator

One-hot encoding converts categorical features into a binary vector with at most a single one value signifying the presence of a specific feature value from among the set of all features.[xi] One-hot encoding categorical variables are a requirement for many machine learning algorithms such as logistic regression and support vector machine. OneHotEncoderEstimator can covert multiple columns, generating a one-hot encoded vector column for each input column.

```
import org.apache.spark.ml.feature.StringIndexer

val df = spark.createDataFrame(
  Seq((0, "Male"), (1, "Male"), (2, "Female"), (3, "Female"),
  (4, "Female"), (5, "Male"))
).toDF("id", "gender")

df.show()

+---+------+
| id|gender|
+---+------+
|  0|  Male|
|  1|  Male|
|  2|Female|
|  3|Female|
|  4|Female|
|  5|  Male|
+---+------+

val indexer = new StringIndexer()
              .setInputCol("gender")
              .setOutputCol("genderIndex")

val indexed = indexer.fit(df).transform(df)

indexed.show()

+---+------+-----------+
| id|gender|genderIndex|
+---+------+-----------+
|  0|  Male|        1.0|
|  1|  Male|        1.0|
|  2|Female|        0.0|
|  3|Female|        0.0|
|  4|Female|        0.0|
|  5|  Male|        1.0|
+---+------+-----------+
```

```
import org.apache.spark.ml.feature.OneHotEncoderEstimator

val encoder = new OneHotEncoderEstimator()
              .setInputCols(Array("genderIndex"))
              .setOutputCols(Array("genderEnc"))

val encoded = encoder.fit(indexed).transform(indexed)

encoded.show()

+---+------+-----------+-------------+
| id|gender|genderIndex|    genderEnc|
+---+------+-----------+-------------+
|  0|  Male|        1.0|    (1,[],[])|
|  1|  Male|        1.0|    (1,[],[])|
|  2|Female|        0.0|(1,[0],[1.0])|
|  3|Female|        0.0|(1,[0],[1.0])|
|  4|Female|        0.0|(1,[0],[1.0])|
|  5|  Male|        1.0|    (1,[],[])|
+---+------+-----------+-------------+
```

SQLTransformer

SQLTransformer allows you to perform data transformation using SQL. A virtual table "__THIS__" corresponds to the input dataset.

```
import org.apache.spark.ml.feature.SQLTransformer

val df = spark.createDataFrame(
  Seq((0, 5.2, 6.7), (2, 25.5, 8.9))).toDF("id", "col1", "col2")

val transformer = new SQLTransformer().setStatement("SELECT ABS(col1 -
col2) as c1, MOD(col1, col2) as c2 FROM __THIS__")

val df2 = transformer.transform(df)

df2.show()
```

```
+----+-----------------+
|  c1|               c2|
+----+-----------------+
| 1.5|              5.2|
|16.6|7.699999999999999|
+----+-----------------+
```

Term Frequency–Inverse Document Frequency (TF–IDF)

TF–IDF or term frequency–inverse document frequency is a feature vectorization method commonly used in text analysis. It is frequently used to indicate the importance of a term or word to a document in the corpus. A transformer, HashingTF, uses feature hashing to convert terms into feature vectors. An estimator, IDF, scales the vectors generated by the HashingTF (or CountVectorizer). I discuss TF–IDF in greater detail in Chapter 4.

```
import org.apache.spark.ml.feature.{HashingTF, IDF, Tokenizer}

val df = spark.createDataFrame(Seq(
  (0, "Kawhi Leonard is the league MVP"),
  (1, "Caravaggio pioneered the Baroque technique"),
  (2, "Using Apache Spark is cool")
)).toDF("label", "sentence")

df.show(false)

+-----+------------------------------------------+
|label|sentence                                  |
+-----+------------------------------------------+
|0    |Kawhi Leonard is the league MVP           |
|1    |Caravaggio pioneered the Baroque technique|
|2    |Using Apache Spark is cool                |
+-----+------------------------------------------+
```

```
val tokenizer = new Tokenizer()
                .setInputCol("sentence")
                .setOutputCol("words")
val df2 = tokenizer.transform(df)

df2.select("label","words").show(false)
+-----+-----------------------------------------------+
|label|words                                          |
+-----+-----------------------------------------------+
|0    |[kawhi, leonard, is, the, league, mvp]         |
|1    |[caravaggio, pioneered, the, baroque, technique]|
|2    |[using, apache, spark, is, cool]               |
+-----+-----------------------------------------------+

val hashingTF = new HashingTF()
                .setInputCol("words")
                .setOutputCol("features")
                .setNumFeatures(20)

val df3 = hashingTF.transform(df2)

df3.select("label","features").show(false)

+-----+-----------------------------------------------+
|label|features                                       |
+-----+-----------------------------------------------+
|0    |(20,[1,4,6,10,11,18],[1.0,1.0,1.0,1.0,1.0,1.0])|
|1    |(20,[1,5,10,12],[1.0,1.0,2.0,1.0])             |
|2    |(20,[1,4,5,15],[1.0,1.0,1.0,2.0])              |
+-----+-----------------------------------------------+
val idf = new IDF()
          .setInputCol("features")
          .setOutputCol("scaledFeatures")

val idfModel = idf.fit(df3)
```

```
val df4 = idfModel.transform(df3)

df4.select("label", "scaledFeatures").show(3,50)
+-----+--------------------------------------------------+
|label|                                    scaledFeatures|
+-----+--------------------------------------------------+
|    0|(20,[1,4,6,10,11,18],[0.0,0.28768207245178085,0...|
|    1|(20,[1,5,10,12],[0.0,0.28768207245178085,0.5753...|
|    2|(20,[1,4,5,15],[0.0,0.28768207245178085,0.28768...|
+-----+--------------------------------------------------+
```

Principal Component Analysis (PCA)

Principal component analysis (PCA) is a dimensionality reduction technique that combines correlated features into a smaller set of linearly uncorrelated features known as principal components. PCA has applications in multiple fields such as image recognition and anomaly detection. I discuss PCA in greater detail in Chapter 4.

```
import org.apache.spark.ml.feature.PCA
import org.apache.spark.ml.linalg.Vectors

val data = Array(
  Vectors.dense(4.2, 5.4, 8.9, 6.7, 9.1),
  Vectors.dense(3.3, 8.2, 7.0, 9.0, 7.2),
  Vectors.dense(6.1, 1.4, 2.2, 4.3, 2.9)
)
val df = spark.createDataFrame(data.map(Tuple1.apply)).toDF("features")

val pca = new PCA()
          .setInputCol("features")
          .setOutputCol("pcaFeatures")
          .setK(2)
          .fit(df)

val result = pca.transform(df).select("pcaFeatures")

result.show(false)
```

```
+--------------------------------------+
|pcaFeatures                           |
+--------------------------------------+
|[13.62324332562565,3.1399510055159445] |
|[14.130156836243236,-1.432033103462711]|
|[3.4900743524527704,0.6866090886347056]|
+--------------------------------------+
```

ChiSqSelector

ChiSqSelector uses the chi-squared independence test for feature selection. The chi-squared test is a method for testing the relationship of two categorical variables. *numTopFeatures* is the default selection method. It returns a set number of features based on chi-squared test, or the features with the most predictive influence. Other selection methods include *percentile, fpr, fdr, and fwe.*

```
import org.apache.spark.ml.feature.ChiSqSelector
import org.apache.spark.ml.linalg.Vectors

val data = Seq(
  (0, Vectors.dense(5.1, 2.9, 5.6, 4.8), 0.0),
  (1, Vectors.dense(7.3, 8.1, 45.2, 7.6), 1.0),
  (2, Vectors.dense(8.2, 12.6, 19.5, 9.21), 1.0)
)

val df = spark.createDataset(data).toDF("id", "features", "class")

val selector = new ChiSqSelector()
               .setNumTopFeatures(1)
               .setFeaturesCol("features")
               .setLabelCol("class")
               .setOutputCol("selectedFeatures")

val df2 = selector.fit(df).transform(df)

df2.show()
```

```
+---+--------------------+-----+---------------+
| id|            features|class|selectedFeatures|
+---+--------------------+-----+---------------+
|  0|   [5.1,2.9,5.6,4.8]|  0.0|          [5.1]|
|  1|  [7.3,8.1,45.2,7.6]|  1.0|          [7.3]|
|  2|[8.2,12.6,19.5,9.21]|  1.0|          [8.2]|
+---+--------------------+-----+---------------+
```

Correlation

Correlation evaluates the strength of the linear relationship between two variables. For linear problems you can use correlation to select relevant features (feature-class correlation) and identify redundant features (intra-feature correlation). Spark MLlib supports both Pearson's and Spearman's correlation. In the following example, correlation calculates the correlation matrix for the input vectors.

```
import org.apache.spark.ml.linalg.{Matrix, Vectors}
import org.apache.spark.ml.stat.Correlation
import org.apache.spark.sql.Row

val data = Seq(
  Vectors.dense(5.1, 7.0, 9.0, 6.0),
  Vectors.dense(3.2, 1.1, 6.0, 9.0),
  Vectors.dense(3.5, 4.2, 9.1, 3.0),
  Vectors.dense(9.1, 2.6, 7.2, 1.8)
)

val df = data.map(Tuple1.apply).toDF("features")

+-----------------+
|         features|
+-----------------+
|[5.1,7.0,9.0,6.0]|
|[3.2,1.1,6.0,9.0]|
|[3.5,4.2,9.1,3.0]|
|[9.1,2.6,7.2,1.8]|
+-----------------+
```

```
val Row(c1: Matrix) = Correlation.corr(df, "features").head

c1: org.apache.spark.ml.linalg.Matrix =
1.0                   -0.01325851107237613  -0.08794286922175912 -0.6536434849076798
-0.01325851107237613  1.0                    0.8773748081826724  -0.1872850762579899
-0.08794286922175912  0.8773748081826724     1.0                 -0.46050932066780714
-0.6536434849076798   -0.1872850762579899   -0.46050932066780714  1.0

val Row(c2: Matrix) = Correlation.corr(df, "features", "spearman").head

c2: org.apache.spark.ml.linalg.Matrix =
1.0                   0.399999999999999     0.19999999999999898  -0.8000000000000014
0.399999999999999     1.0                   0.8000000000000035   -0.19999999999999743
0.19999999999999898   0.8000000000000035    1.0                  -0.39999999999999486
-0.8000000000000014   -0.19999999999999743  -0.39999999999999486  1.0
```

You can also calculate the correlation of values stored in DataFrame columns as shown in the following.

```
dataDF.show

+------------+-----------+------------+-----------+-----------+-----+
|sepal_length|sepal_width|petal_length|petal_width|      class|label|
+------------+-----------+------------+-----------+-----------+-----+
|         5.1|        3.5|         1.4|        0.2|Iris-setosa|  0.0|
|         4.9|        3.0|         1.4|        0.2|Iris-setosa|  0.0|
|         4.7|        3.2|         1.3|        0.2|Iris-setosa|  0.0|
|         4.6|        3.1|         1.5|        0.2|Iris-setosa|  0.0|
|         5.0|        3.6|         1.4|        0.2|Iris-setosa|  0.0|
|         5.4|        3.9|         1.7|        0.4|Iris-setosa|  0.0|
|         4.6|        3.4|         1.4|        0.3|Iris-setosa|  0.0|
|         5.0|        3.4|         1.5|        0.2|Iris-setosa|  0.0|
|         4.4|        2.9|         1.4|        0.2|Iris-setosa|  0.0|
|         4.9|        3.1|         1.5|        0.1|Iris-setosa|  0.0|
|         5.4|        3.7|         1.5|        0.2|Iris-setosa|  0.0|
|         4.8|        3.4|         1.6|        0.2|Iris-setosa|  0.0|
|         4.8|        3.0|         1.4|        0.1|Iris-setosa|  0.0|
|         4.3|        3.0|         1.1|        0.1|Iris-setosa|  0.0|
|         5.8|        4.0|         1.2|        0.2|Iris-setosa|  0.0|
```

```
|         5.7|        4.4|         1.5|        0.4|Iris-setosa|  0.0|
|         5.4|        3.9|         1.3|        0.4|Iris-setosa|  0.0|
|         5.1|        3.5|         1.4|        0.3|Iris-setosa|  0.0|
|         5.7|        3.8|         1.7|        0.3|Iris-setosa|  0.0|
|         5.1|        3.8|         1.5|        0.3|Iris-setosa|  0.0|
+------------+-----------+------------+-----------+-----------+-----+

dataDF.stat.corr("petal_length","label")
res48: Double = 0.9490425448523336

dataDF.stat.corr("petal_width","label")
res49: Double = 0.9564638238016178

dataDF.stat.corr("sepal_length","label")
res50: Double = 0.7825612318100821

dataDF.stat.corr("sepal_width","label")
res51: Double = -0.41944620026002677
```

Evaluation Metrics

As discussed in Chapter 1, precision, recall, and accuracy are important evaluation metrics for evaluating a model's performance. However, they may not always be the best metrics for certain problems.

Area Under the Receiver Operating Characteristic (AUROC)

The area under the receiver operating characteristic (AUROC) is a common performance metric for evaluating binary classifiers. The receiver operating characteristic (ROC) is a graph that plots the true positive rate against the false positive rate. The area under the curve (AUC) is the area below the ROC curve. The AUC can be interpreted as the probability that the model ranks a random positive example higher than a random negative example.[xii] The larger the area under the curve (the closer the AUROC is to 1.0), the better the model is performing. A model with AUROC of 0.5 is useless since its predictive accuracy is just as good as random guessing.

```
import org.apache.spark.ml.evaluation.BinaryClassificationEvaluator

val evaluator = new BinaryClassificationEvaluator()
                .setMetricName("areaUnderROC")
                .setRawPredictionCol("rawPrediction")
                .setLabelCol("label")
```

F1 Measure

The F1 measure or F1 score is the harmonic mean or weighted average of precision and recall. It is a common performance metric for evaluating multiclass classifiers. It is also a good measure when there is an uneven class distribution. The best F1 score is 1, while the worst score is 0. A good F1 measure means that you have low false negatives and low false positives. The formula for F1 measure is: *F1-Measure = 2 ∗ (precision ∗ recall) / (precision + recall).*

```
import org.apache.spark.ml.evaluation.MulticlassClassificationEvaluator

val evaluator = new MulticlassClassificationEvaluator()
                .setMetricName("f1")
                .setLabelCol("label")
                .setPredictionCol("prediction")
```

Root Mean Squared Error (RMSE)

The root mean squared error (RMSE) is the most common metric for regression tasks. The RMSE is simply the square root of the mean square error (MSE). The MSE indicates how close a regression line to a set of data points by taking the distances or "errors" from the points to the regression line and squaring them.[xiii] The smaller the MSE, the better the fit. However, the MSE does not match the unit of the original data since the value is squared. The RMSE has the same unit as the output.

```
import org.apache.spark.ml.evaluation.RegressionEvaluator

val evaluator = new RegressionEvaluator()
                .setLabelCol("label")
                .setPredictionCol("prediction")
                .setMetricName("rmse")
```

I cover other evaluation metrics such as Within Set Sum of Squared Errors (WSSSE) and silhouette coefficient in subsequent chapters. For a complete list of all the evaluation metrics supported by Spark MLlib, consult Spark's online documentation.

Model Persistence

Spark MLlib allows you to save models and load them at a later time. This is particularly useful if you want to integrate your model with third-party applications or share them with other members of your team.

Saving a single Random Forest model

```
rf = RandomForestClassifier(numBin=10,numTrees=30)
model = rf.fit(training)
model.save("modelpath")
```

Loading a single Random Forest model

```
val model2 = RandomForestClassificationModel.load("modelpath")
```

Saving a full pipeline

```
val pipeline = new Pipeline().setStages(Array(labelIndexer,vectorAssembler,
rf))
val cv = new CrossValidator().setEstimator(pipeline)
val model = cv.fit(training)
model.save("modelpath")
```

Loading a full pipeline

```
val model2 = CrossValidatorModel.load("modelpath")
```

Spark MLlib Example

Let's work on an example. We'll use the Heart Disease Data Set[xiv] from the UCI Machine Learning Repository to predict the presence of heart disease. The data was collected by Robert Detrano, MD, PhD, and his team at the VA Medical Center, Long Beach and Cleveland Clinic Foundation. Historically, the Cleveland dataset has been the subject of numerous studies, so we'll use that dataset. The original dataset has 76 attributes, but only 14 of them are used in ML research (Table 2-1). We will perform binomial classification and determine if the patient has a heart disease or not (Listing 2-2).

Table 2-1. *Cleveland Heart Disease Data Set Attribute Information*

Attribute	Description
age	Age
sex	Sex
cp	Chest pain type
trestbps	Resting blood pressure
chol	Serum cholesterol in mg/dl
fbs	Fasting blood sugar > 120 mg/dl
restecg	Resting electrocardiographic results
thalach	Maximum heart rate achieved
exang	Exercise-induced angina
oldpeak	ST depression induced by exercise relative to rest
slope	The slope of the peak exercise ST segment
ca	Number of major vessels (0–3) colored by flourosopy
thal	Thallium stress test result
num	The predicted attribute – diagnosis of heart disease

Let's start. Download the file and copy it to HDFS.

```
wget http://archive.ics.uci.edu/ml/machine-learning-databases/heart-
disease/cleveland.data
```

```
head -n 10 processed.cleveland.data
```

```
63.0,1.0,1.0,145.0,233.0,1.0,2.0,150.0,0.0,2.3,3.0,0.0,6.0,0
67.0,1.0,4.0,160.0,286.0,0.0,2.0,108.0,1.0,1.5,2.0,3.0,3.0,2
67.0,1.0,4.0,120.0,229.0,0.0,2.0,129.0,1.0,2.6,2.0,2.0,7.0,1
37.0,1.0,3.0,130.0,250.0,0.0,0.0,187.0,0.0,3.5,3.0,0.0,3.0,0
41.0,0.0,2.0,130.0,204.0,0.0,2.0,172.0,0.0,1.4,1.0,0.0,3.0,0
56.0,1.0,2.0,120.0,236.0,0.0,0.0,178.0,0.0,0.8,1.0,0.0,3.0,0
```

```
62.0,0.0,4.0,140.0,268.0,0.0,2.0,160.0,0.0,3.6,3.0,2.0,3.0,3
57.0,0.0,4.0,120.0,354.0,0.0,0.0,163.0,1.0,0.6,1.0,0.0,3.0,0
63.0,1.0,4.0,130.0,254.0,0.0,2.0,147.0,0.0,1.4,2.0,1.0,7.0,2
53.0,1.0,4.0,140.0,203.0,1.0,2.0,155.0,1.0,3.1,3.0,0.0,7.0,1
```

```
hadoop fs -put processed.cleveland.data /tmp/data
```

We use spark-shell to interactively train our model.

Listing 2-2. Performing Binary Classification Using Random Forest

```
spark-shell

val dataDF = spark.read.format("csv")
             .option("header", "true")
             .option("inferSchema", "true")
             .load(d("/tmp/data/processed.cleveland.data")
             .toDF("id","age","sex","cp","trestbps","chol","fbs","restecg",
             "thalach","exang","oldpeak","slope","ca","thal","num")

dataDF.printSchema
root
 |-- id: string (nullable = false)
 |-- age: float (nullable = true)
 |-- sex: float (nullable = true)
 |-- cp: float (nullable = true)
 |-- trestbps: float (nullable = true)
 |-- chol: float (nullable = true)
 |-- fbs: float (nullable = true)
 |-- restecg: float (nullable = true)
 |-- thalach: float (nullable = true)
 |-- exang: float (nullable = true)
 |-- oldpeak: float (nullable = true)
 |-- slope: float (nullable = true)
 |-- ca: float (nullable = true)
 |-- thal: float (nullable = true)
 |-- num: float (nullable = true)
```

```
val myFeatures = Array("age", "sex", "cp", "trestbps", "chol", "fbs",
      "restecg", "thalach", "exang", "oldpeak", "slope",
      "ca", "thal", "num")

import org.apache.spark.ml.feature.VectorAssembler

val assembler = new VectorAssembler()
                .setInputCols(myFeatures)
                .setOutputCol("features")

val dataDF2 = assembler.transform(dataDF)

import org.apache.spark.ml.feature.StringIndexer

val labelIndexer = new StringIndexer()
                   .setInputCol("num")
                   .setOutputCol("label")

val dataDF3 = labelIndexer.fit(dataDF2).transform(dataDF2)


val dataDF4 = dataDF3.where(dataDF3("ca").isNotNull)
              .where(dataDF3("thal").isNotNull)
              .where(dataDF3("num").isNotNull)

val Array(trainingData, testData) = dataDF4.randomSplit(Array(0.8, 0.2), 101)
import org.apache.spark.ml.classification.RandomForestClassifier

val rf = new RandomForestClassifier()
         .setFeatureSubsetStrategy("auto")
         .setSeed(101)

import org.apache.spark.ml.evaluation.BinaryClassificationEvaluator

val evaluator = new BinaryClassificationEvaluator().setLabelCol("label")

import org.apache.spark.ml.tuning.ParamGridBuilder

val pgrid = new ParamGridBuilder()
      .addGrid(rf.maxBins, Array(10, 20, 30))
      .addGrid(rf.maxDepth, Array(5, 10, 15))
```

```
      .addGrid(rf.numTrees, Array(20, 30, 40))
      .addGrid(rf.impurity, Array("gini", "entropy"))
      .build()

import org.apache.spark.ml.Pipeline

val pipeline = new Pipeline().setStages(Array(rf))

import org.apache.spark.ml.tuning.CrossValidator

val cv = new CrossValidator()
      .setEstimator(pipeline)
      .setEvaluator(evaluator)
      .setEstimatorParamMaps(pgrid)
      .setNumFolds(3)
```

We can now fit the model.

```
val model = cv.fit(trainingData)
```

Perform predictions on test data.

```
val prediction = model.transform(testData)
```

Let's evaluate the model.

```
import org.apache.spark.ml.param.ParamMap

val pm = ParamMap(evaluator.metricName -> "areaUnderROC")

val aucTestData = evaluator.evaluate(prediction, pm)
```

Graph Processing

Spark includes a graph processing framework called GraphX. There is a separate package called GraphFrames which is based on DataFrames. GraphFrames is currently not part of core Apache Spark. GraphX and GraphFrames are still in active development at the time of this writing.[xv] I cover GraphX in Chapter 6.

Beyond Spark MLlib: Third-Party Machine Learning Integrations

Spark has access to a rich ecosystem of third-party frameworks and libraries, thanks to countless open source contributors as well as companies such as Microsoft and Google. While I cover core Spark MLlib algorithms, this book is focused on more powerful, next-generation algorithms and frameworks such as XGBoost, LightGBM, Isolation Forest, Spark NLP, and distributed deep learning. I will cover them in the succeeding chapters.

Optimizing Spark and Spark MLlib with Alluxio

Alluxio, formerly known as Tachyon, is an open source project from UC Berkeley AMPLab. Alluxio is a distributed memory-centric storage system originally developed as a research project by Haoyuan Li in 2012, then a PhD student and a founding Apache Spark committer at AMPLab.[xvi] The project is the storage layer of the Berkeley Data Analytics Stack (BDAS). In 2015, Alluxio, Inc. was founded by Li to commercialize Alluxio, receiving a $7.5 million cash infusion from Andreessen Horowitz. Today, Alluxio has more than 200 contributors from 50 organizations around the world such as Intel, IBM, Yahoo, and Red Hat. Several high-profile companies are currently using Alluxio in production such as Baidu, Alibaba, Rackspace, and Barclays.[xvii]

Alluxio can be used to optimize Spark machine learning and deep learning workloads by enabling ultrafast big data storage to extremely large datasets. Deep learning benchmarks conducted by Alluxio show significant performance improvements when reading data from Alluxio instead of S3.[xviii]

Architecture

Alluxio is a memory-centric distributed storage system and aims to be the de facto storage unification layer for big data. It provides a virtualization layer that unifies access for different storage engines such as Local FS, HDFS, S3, and NFS and computing frameworks such as Spark, MapReduce, Hive, and Presto. Figure 2-4 gives you an overview of Alluxio's architecture.

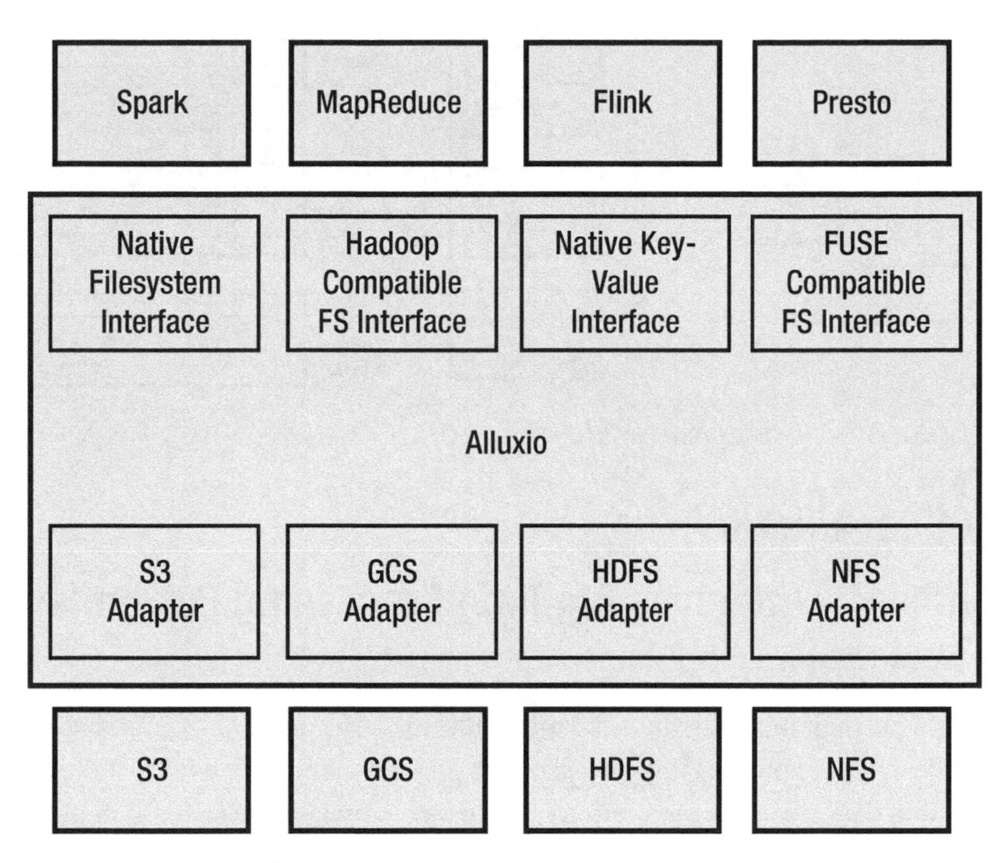

Figure 2-4. *Alluxio architecture overview*

Alluxio is the middle layer that coordinates data sharing and directs data access while at the same time providing computing frameworks and big data applications high-performance low-latency memory speed. Alluxio integrates seamlessly with Spark and Hadoop, only requiring minor configuration changes. By taking advantage of Alluxio's unified namespace feature, applications only need to connect to Alluxio to access data stored in any of the supported storage engines. Alluxio has its own native API as well as a Hadoop-compatible file system interface. The convenience class enables users to execute code originally written for Hadoop without any code changes. A REST API provides access to other languages. We will explore the APIs later in the chapter.

Alluxio's unified namespace feature does not support relational databases and MPP engines such as Redshift or Snowflake or document databases such as MongoDB. Of course, writing to and from Alluxio and the storage engines mentioned is supported. Developers can use a computing framework such as Spark to create a data frame from a Redshift table and store it in an Alluxio file system in Parquet or CSV format, and vice versa (Figure 2-5).

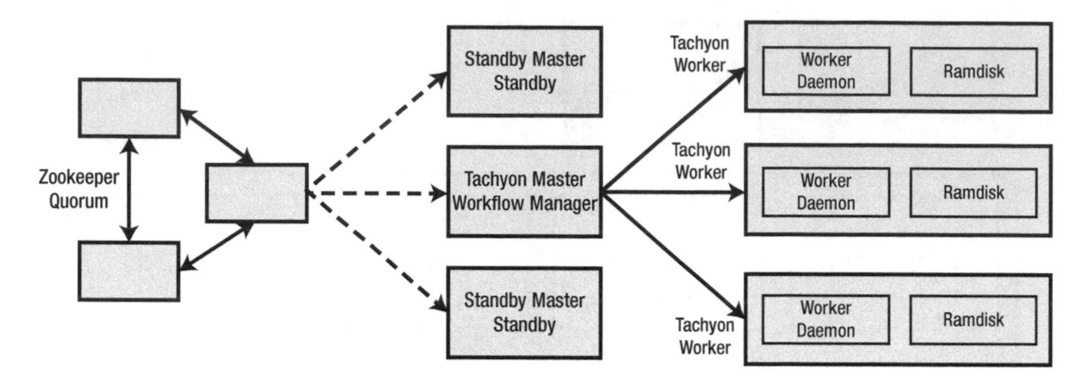

Figure 2-5. *Alluxio technical architecture*

Why Use Alluxio?

Significantly Improves Big Data Processing Performance and Scalability

Over the years, memory has gotten cheaper, while its performance has gotten faster. Meanwhile, performance of hard drives has only gotten marginally better. There is no question that data processing in memory is an order of magnitude faster than processing data on disk. In almost all programming paradigms, we are advised to cache data in memory to improve performance. One of the main advantages of Apache Spark over MapReduce is its ability to cache data. Alluxio takes that to the next level, providing big data applications not just a caching layer, but a full-blown distributed high-performance memory-centric storage system.

Baidu is operating one of the largest Alluxio clusters in the world, with 1000 worker nodes handling more than 2PB of data. With Alluxio, Baidu is seeing an average of 10x and up to 30x performance improvement in query and processing time, significantly improving Baidu's ability to make important business decisions.[xix] Barclays published an article describing their experience with Alluxio. Barclays Data Scientist Gianmario Spacagna and Harry Powell, Head of Advanced Analytics, were able to tune their Spark jobs from hours to seconds using Alluxio.[xx] Qunar.com, one of China's largest travel search engine, experienced a 15x–300x performance improvement using Alluxio.[xxi]

Multiple Frameworks and Applications Can Share Data at Memory Speed

A typical big data cluster has multiple sessions running different computing frameworks such as Spark and MapReduce. In case of Spark, each application gets its own executor processes, with each task within an executor running on its own JVM, isolating Spark applications from each other. This means that Spark (and MapReduce) applications have no way of sharing data, except writing to a storage system such as HDFS or S3. As shown in Figure 2-6, a Spark job and a MapReduce job are using the same data stored in HDFS or S3. In Figure 2-7, multiple Spark jobs are using the same data with each job storing its own version of the data in its own heap space.[xxii] Not only is data duplicated but sharing data via HDFS or S3 can be slow, particularly if you're sharing large amounts of data.

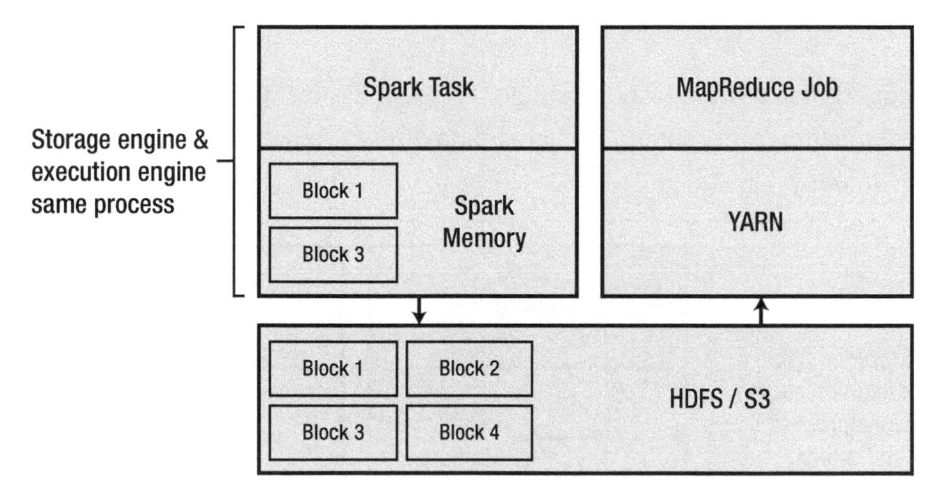

Figure 2-6. *Different frameworks sharing data via HDFS or S3*

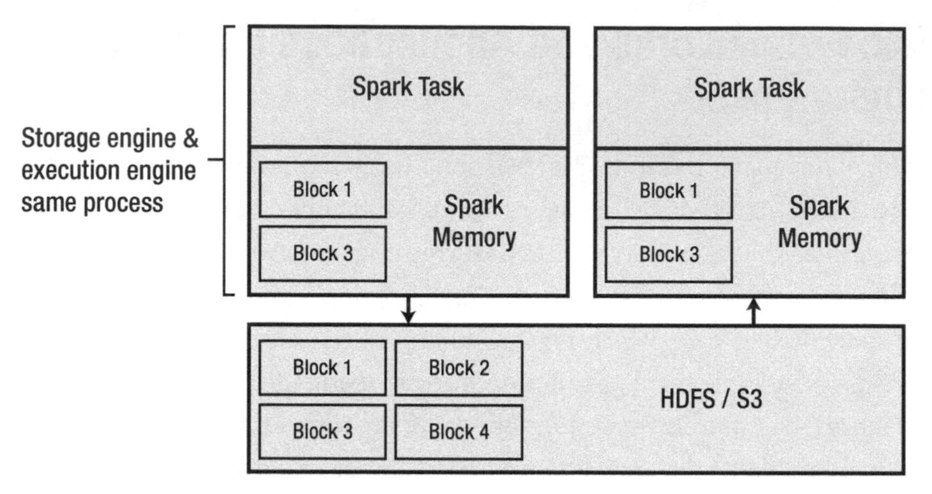

Figure 2-7. *Different jobs sharing data via HDFS or S3*

By using Alluxio as an off-heap storage (Figure 2-8), multiple frameworks and jobs can share data at memory speed, reducing data duplication, increasing throughput, and decreasing latency.

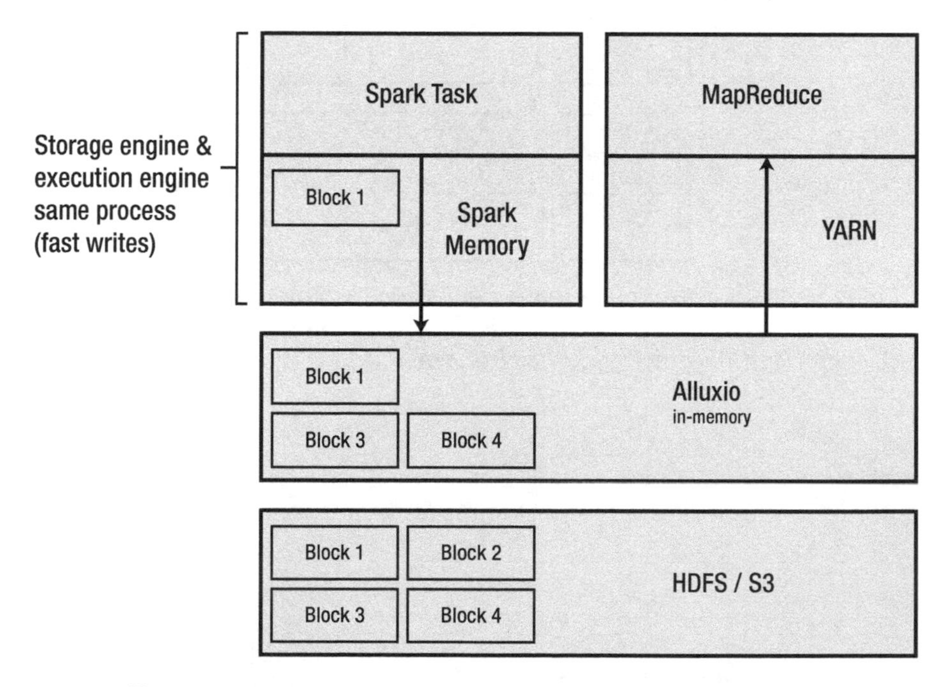

Figure 2-8. *Different jobs and frameworks sharing data at memory speed*

Provides High Availability and Persistence in Case of Application Termination or Failure

In Spark, the executor processes and the executor memory resides in the same JVM, with all cached data stored in the JVM heap space (Figure 2-9).

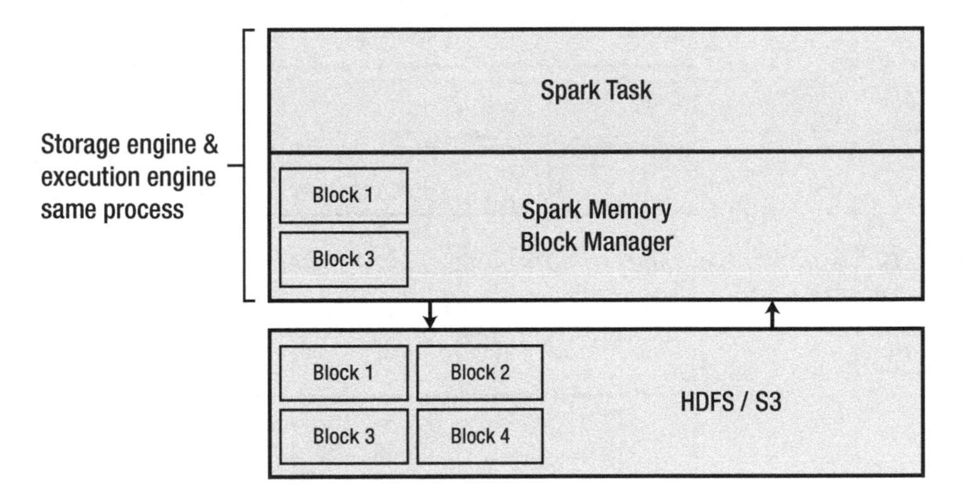

Figure 2-9. *Spark job with its own heap memory*

When the job completes or for some reason the JVM crashes due to runtime exceptions, all the data cached in heap space will be lost as shown in Figures 2-10 and 2-11.

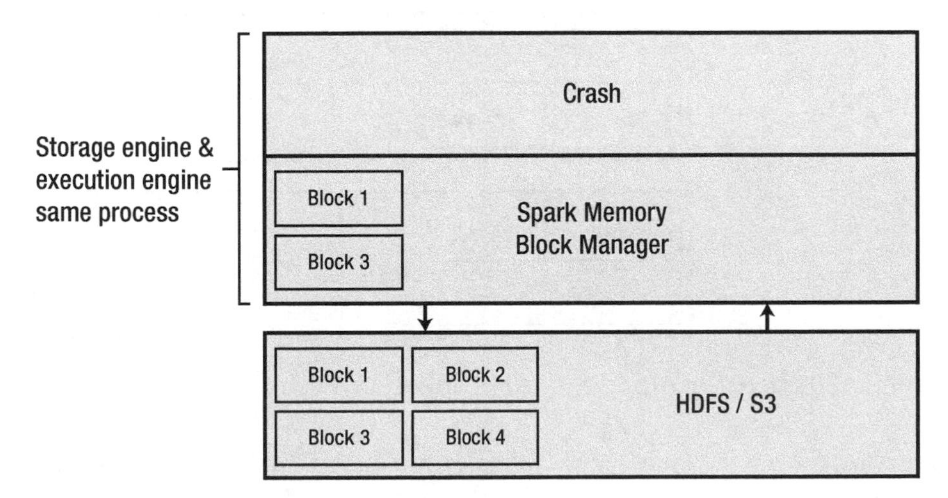

Figure 2-10. *Spark job crashes or completes*

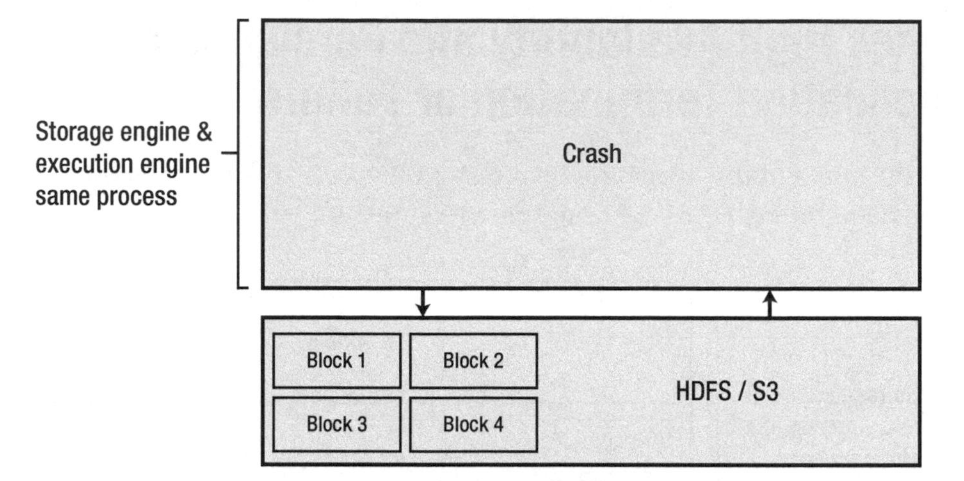

Figure 2-11. *Spark job crashes or completes. Heap space is lost*

The solution is to use Alluxio as an off-heap storage (Figure 2-12).

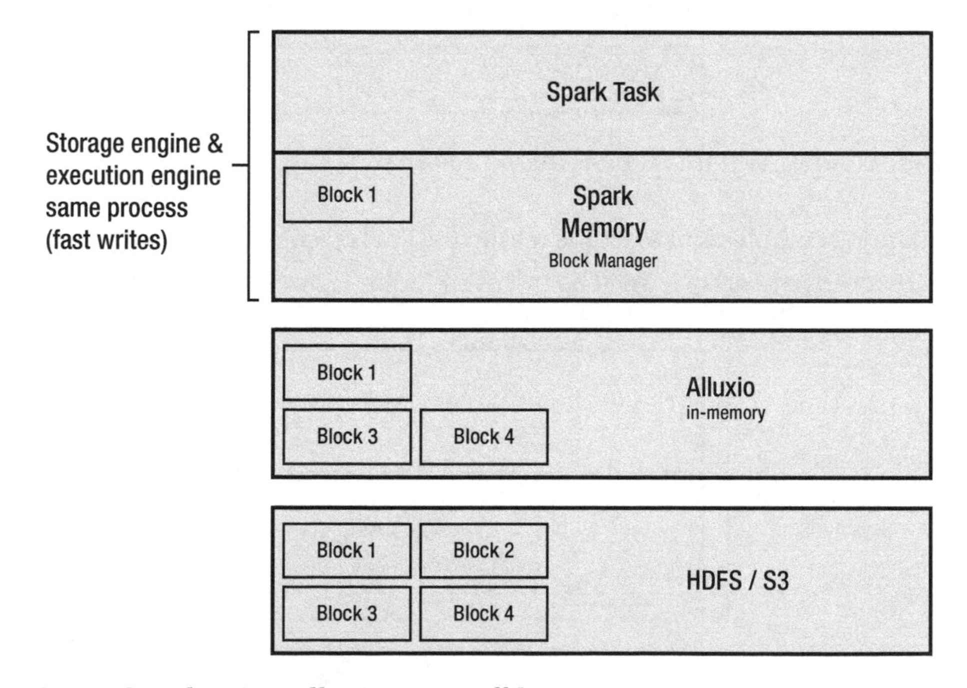

Figure 2-12. *Spark using Alluxio as an off-heap storage*

In this case, even if the Spark JVM crashes, the data is still available in Alluxio (Figures 2-13 and 2-14).

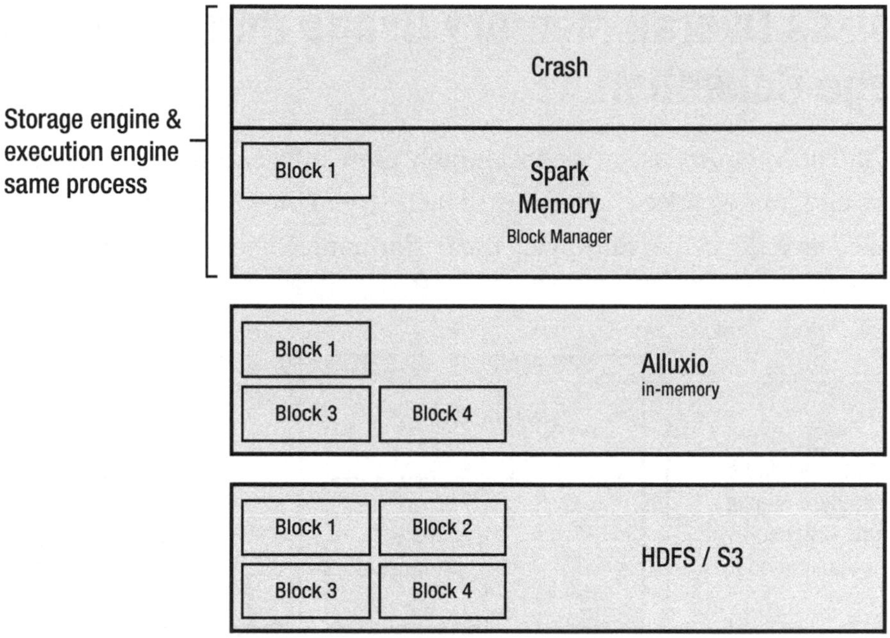

Storage engine &
execution engine
same process
Crash
Block 1
Spark
Memory
Block Manager
Block 1
Block 3
Block 4
Alluxio
in-memory
Block 1
Block 2
Block 3
Block 4
HDFS / S3

Figure 2-13. *Spark job crashes or completes*

Storage engine &
execution engine
same process
Crash
Block 1
Block 3
Block 4
Alluxio
in-memory
Block 1
Block 2
Block 3
Block 4
HDFS / S3

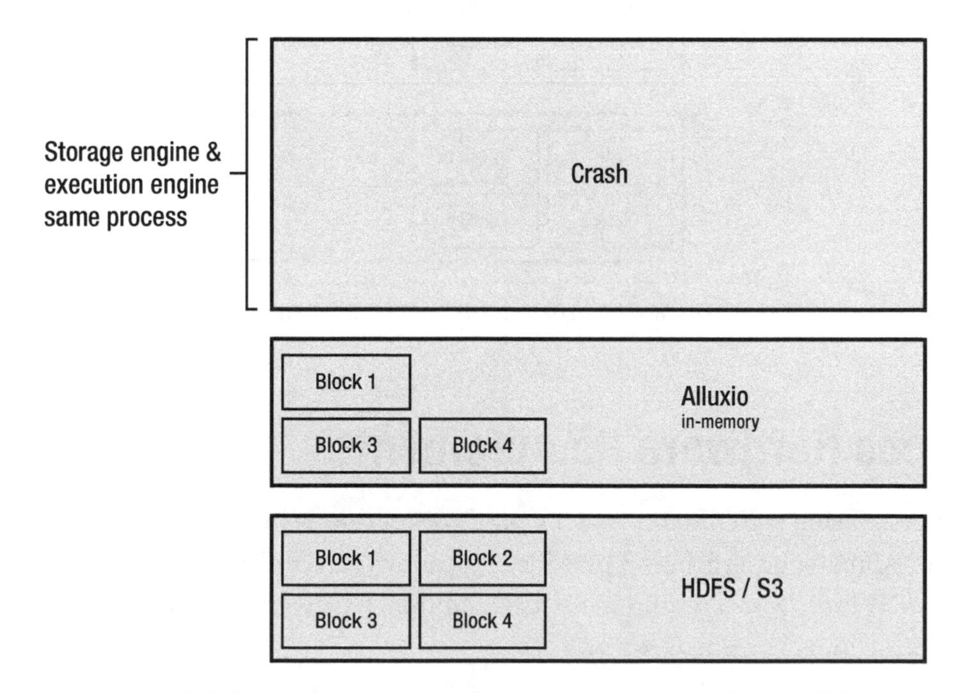

Figure 2-14. *Spark job crashes or completes. Heap space is lost. Off-heap memory is still available*

Optimizes Overall Memory Usage and Minimizes Garbage Collection

By using Alluxio, memory usage is considerably more efficient since data is shared across jobs and frameworks, and because data is stored off-heap, garbage collection is minimized as well, further improving the performance of jobs and applications (Figure 2-15).

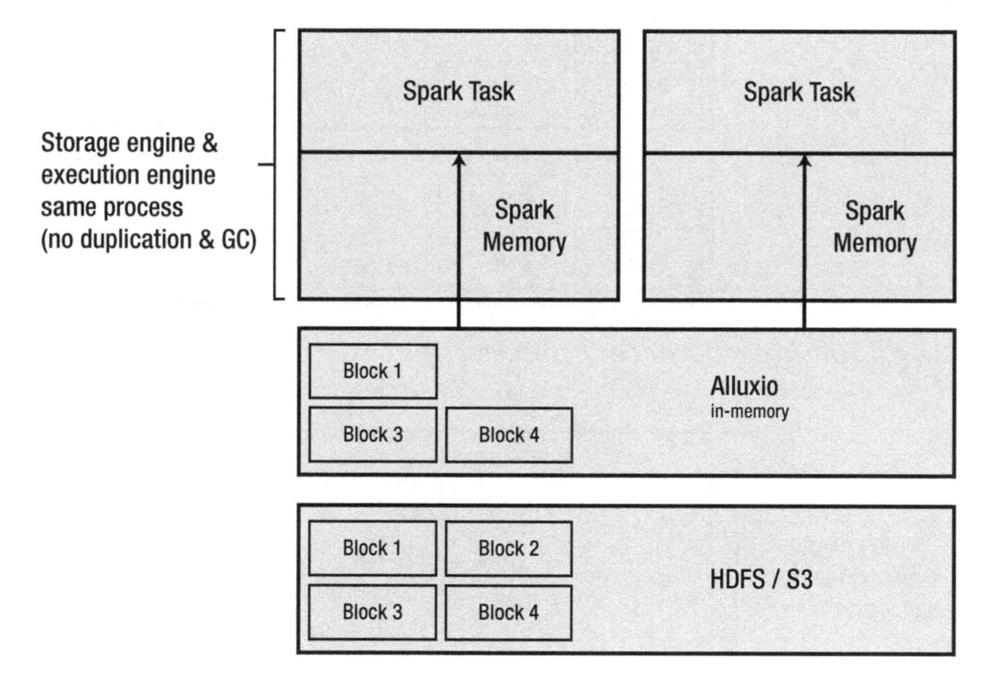

Figure 2-15. *Multiple Spark and MapReduce jobs can access the same data stored in Alluxio*

Reduces Hardware Requirements

Big data processing with Alluxio is significantly faster than with HDFS and S3. IBM's tests show Alluxio outperforming HDFS by 110x for write IO.[xxiii] With that kind of performance, there is less requirement for additional hardware, thus saving you in infrastructure and licensing cost.

Apache Spark and Alluxio

You access data in Alluxio similar to how you would access data stored in HDFS and S3 from Spark.

```
val dataRDD = sc.textFile("alluxio://localhost:19998/test01.csv")

val parsedRDD = dataRDD.map{_.split(",")}

case class CustomerData(userid: Long, city: String, state: String,
age: Short)

val dataDF = parsedRDD.map{ a =>CustomerData(a(0).toLong, a(1).toString,
a(2).toString, a(3).toShort) }.toDF

dataDF.show()

+------+---------------+-----+---+
|userid|           city|state|age|
+------+---------------+-----+---+
|   300|       Torrance|   CA| 23|
|   302|Manhattan Beach|   CA| 21|
+------+---------------+-----+---+
```

Summary

This chapter gave you a quick introduction to Spark and Spark MLlib, enough to give you the skills needed to perform common data processing and machine learning tasks. My goal was to get you up to speed as fast as possible. For a more thorough treatment, *Spark: The Definitive Guide* by Bill Chambers and Matei Zaharia (O'Reilly, 2018) provides a thorough introduction to Spark. *Scala Programming for Big Data Analytics* by Irfan Elahi (Apress, 2019), *Learning Scala* by Jason Swartz (O'Reilly, 2014), and *Programming in Scala* by Martin Odersky, Lex Spoon, and Bill Venners (Artima, 2016) are good introduction to Scala. I also gave an introduction to Alluxio, an in-memory distributed computing platform that can be used to optimize large-scale machine learning and deep learning workloads.

References

i. Peter Norvig et al.; "The Unreasonable Effectiveness of Data," googleuserconent.com, 2009, https://static.googleusercontent.com/media/research.google.com/en//pubs/archive/35179.pdf

ii. Spark; "Spark Overview," spark.apache.org, 2019, https://spark.apache.org/docs/2.2.0/

iii. Apache Software Foundation; "The Apache Software Foundation Announces Apache Spark as a Top-Level Project," blogs.apache.org, 2014, https://blogs.apache.org/foundation/entry/the_apache_software_foundation_announces50

iv. Spark; "Spark News," spark.apache.org, 2019, https://spark.apache.org/news/

v. Reddit; "Matei Zaharia AMA," reddit.com, 2015, www.reddit.com/r/IAmA/comments/31bkue/im_matei_zaharia_creator_of_spark_and_cto_at/?st=j1svbrx9&sh=a8b9698e

vi. Databricks; "Apache Spark," Databricks.com, 2019, https://databricks.com/spark/about

vii. Databricks; "How to use SparkSession in Apache Spark 2.0," Databricks.com, 2016, https://databricks.com/blog/2016/08/15/how-to-use-sparksession-in-apache-spark-2-0.html

viii. Solr; "Using SolrJ," lucene.apache.org, 2019, https://lucene.apache.org/solr/guide/6_6/using-solrj.html

ix. Lucidworks; "Lucidworks Spark/Solr Integration," github.com, 2019, https://github.com/lucidworks/spark-solr

x. Spark; "Machine Learning Library (MLlib) Guide," spark.apache.org, http://spark.apache.org/docs/latest/ml-guide.html

xi. Spark; "OneHotEncoderEstimator," spark.apache.org, 2019, https://spark.apache.org/docs/latest/ml-features#onehotencoderestimator

xii. Google; "Classification: ROC Curve and AUC," developers.google.com, 2019, https://developers.google.com/machine-learning/crash-course/classification/roc-and-auc

xiii. Stephanie Glen; "Mean Squared Error: Definition and Example," statisticshowto.datasciencecentral.com, 2013, www.statisticshowto.datasciencecentral.com/mean-squared-error/

xiv. Andras Janosi, William Steinbrunn, Matthias Pfisterer, Robert Detrano; "Heart Disease Data Set," archive.ics.uci.edu, 1988, http://archive.ics.uci.edu/ml/datasets/heart+Disease

xv. Spark; "GraphX," spark.apache.org, 2019, https://spark.apache.org/graphx/

xvi. Chris Mattman; "Apache Spark for the Incubator," mail-archives.apache.org, 2013, http://mail-archives.apache.org/mod_mbox/incubator-general/201306.mbox/%3CCDD80F64.D5F9D%25chris.a.mattmann@jpl.nasa.gov%3E

xvii. Haoyuan Li; "Alluxio, formerly Tachyon, is Entering a New Era with 1.0 release," alluxio.io, 2016, www.alluxio.com/blog/alluxio-formerly-tachyon-is-entering-a-new-era-with-10-release

xviii. Yupeng Fu; "Flexible and Fast Storage for Deep Learning with Alluxio," alluxio.io, 2018, www.alluxio.io/blog/flexible-and-fast-storage-for-deep-learning-with-alluxio/

xix. Alluxio; "Alluxio Virtualizes Distributed Storage for Petabyte Scale Computing at In-Memory Speeds," globenewswire.com, 2016, www.marketwired.com/press-release/alluxio-virtualizes-distributed-storage-petabyte-scale-computing-in-memory-speeds-2099053.html

xx. Henry Powell and Gianmario Spacagna; "Making the Impossible Possible with Tachyon: Accelerate Spark Jobs from Hours to Seconds," dzone.com, 2016, https://dzone.com/articles/Accelerate-In-Memory-Processing-with-Spark-from-Hours-to-Seconds-With-Tachyon

xxi. Haoyuan Li; "Alluxio Keynote at Strata+Hadoop World Beijing 2016," slideshare.net, 2016, www.slideshare.net/Alluxio/alluxio-keynote-at-stratahadoop-world-beijing-2016-65172341

xxii. Mingfei S.; "Getting Started with Tachyon by Use Cases," intel.com, 2016, https://software.intel.com/en-us/blogs/2016/02/04/getting-started-with-tachyon-by-use-cases

xxiii. Gil Vernik; "Tachyon for ultra-fast Big Data processing," ibm.com, 2015, www.ibm.com/blogs/research/2015/08/tachyon-for-ultra-fast-big-data-processing/

CHAPTER 3

Supervised Learning

The surest kind of knowledge is what you construct yourself.

—Judea Pearl[i]

Supervised learning is a machine learning task that makes prediction using a training dataset. Supervised learning can be categorized into either classification or regression. Regression is for predicting continuous values such as price, temperature, or distance, while classification is for predicting categories such as yes or no, spam or not spam, or malignant or benign.

Classification

Classification is perhaps the most common supervised machine learning task. You most likely have already encountered applications that utilized classification without even realizing it. Popular use cases include medical diagnosis, targeted marketing, spam detection, credit risk prediction, and sentiment analysis, to mention a few. There are three types of classification tasks.

Binary Classification

A task is binary or binomial classification if there are only two categories. For example, when using binary classification algorithm for spam detection, the output variable can have two categories: spam or not spam. For detecting cancer, the categories can be malignant or benign. For targeted marketing, predicting the likelihood of someone buying an item such as milk, the categories can simply be yes or no.

B. Quinto, *Next-Generation Machine Learning with Spark*, https://doi.org/10.1007/978-1-4842-5669-5_3

Multiclass Classification

Multiclass or multinomial classification tasks have three or more categories. For example, to predict weather conditions you might have five categories: rainy, cloudy, sunny, snowy, and windy. To extend our targeted marketing example, multiclass classification can be used to predict if a customer is more likely to buy whole milk, reduced-fat milk, low-fat milk, or skim milk.

Multilabel Classification

In multilabel classification, multiple categories can be assigned to each observation. In contrast, only one category can be assigned to an observation in multiclass classification. Using our targeted marketing example, multilabel classification is used not only to predict if a customer is more likely to buy milk, but other items as well such as cookies, butter, hotdogs, or bread.

Spark MLlib Classification Algorithms

Spark MLlib includes several algorithms for classification. I will discuss the most popular algorithms and provide easy-to-follow code examples that build upon what we covered in Chapter 2. Later in the chapter, I will discuss more advanced, next-generation algorithms such as XGBoost and LightGBM.

Logistic Regression

Logistic regression is a linear classifier that predicts probabilities. It uses a logistic (sigmoid) function to transform its output into a probability value that can be mapped to two (binary) classes. Multiclass classification is supported through multinomial logistic (softmax) regression.[ii] We will use logistic regression in one of our examples later in the chapter.

Support Vector Machine

Support vector machine is a popular algorithm that works by finding the optimal hyperplane that maximizes the margin between two classes, dividing the data points into separate classes by as wide a gap as possible. The data points closest to the classification boundary are known as support vectors (Figure 3-1).

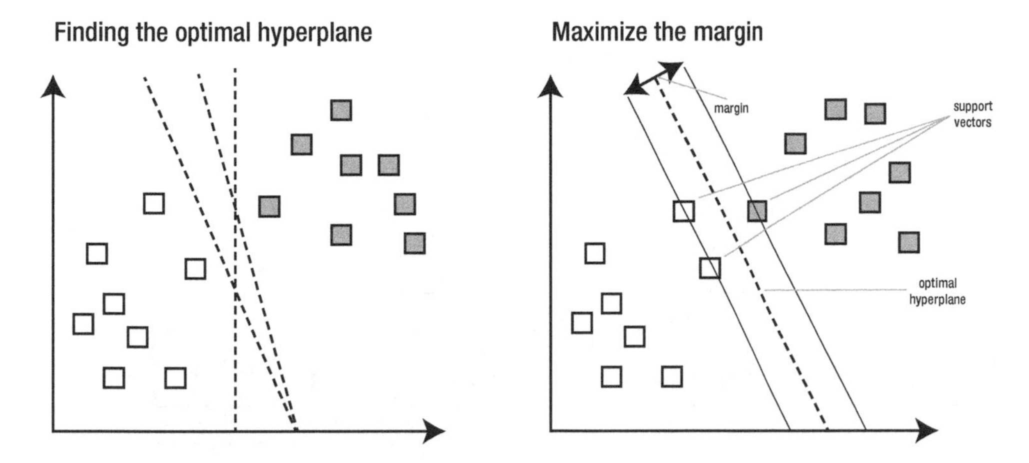

Figure 3-1. *Finding the optimal hyperplane that maximizes the margin between two classes*[iii]

Naïve Bayes

Naïve Bayes is a simple multiclass linear classification algorithm based on Bayes' theorem. Naïve Bayes got its name because it naively assumes that the features in a dataset are independent, ignoring any possible correlations between features. We use naïve Bayes in our sentiment analysis example later in the chapter.

Multilayer Perceptron

Multilayer perceptron is a feedforward artificial network that consists of several fully connected layers of nodes. Nodes in the input layer correspond to the input dataset. Nodes in the intermediate layers utilize a logistic (sigmoid) function, while nodes in the final output layer use a softmax function to support multiclass classification. The number of nodes in the output layer must match the number of classes.[iv] I discuss multiplayer perceptron in Chapter 7.

Decision Trees

A decision tree predicts the value of an output variable by learning decision rules inferred from the input variables.

Visually, a decision tree looks like a tree inverted with the root node at the top. Every internal node represents a test on an attribute. Leaf nodes represent a class label, while an individual branch represents the result of a test. Figure 3-2 shows a decision tree for predicting credit risk.

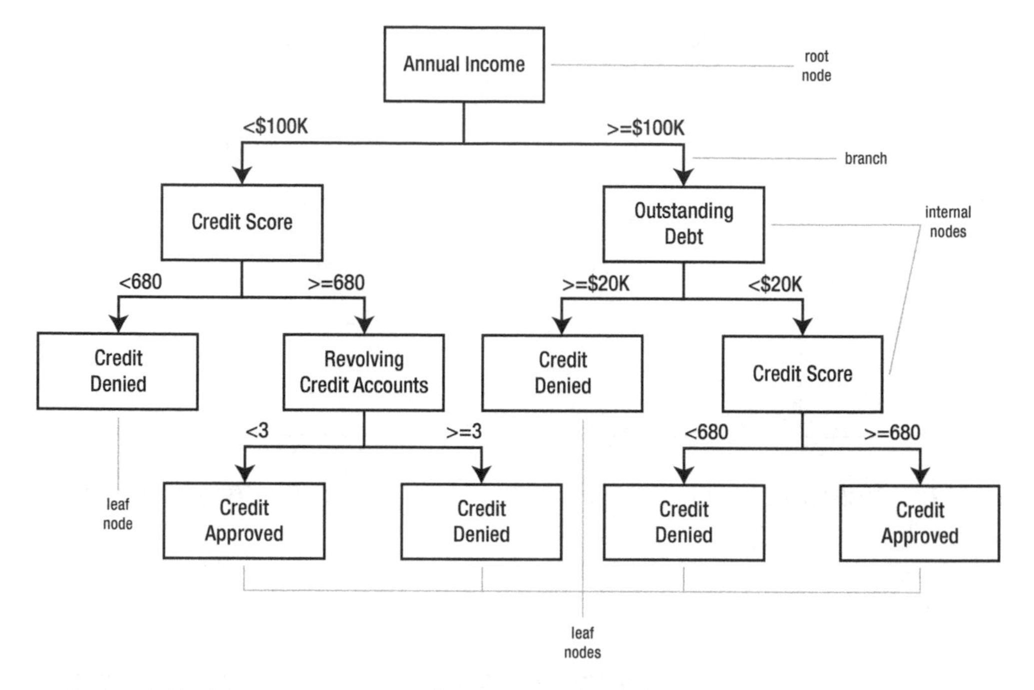

Figure 3-2. *A decision tree for predicting credit risk*

Decision trees perform recursive binary splitting of the feature space. To maximize the information gain, the split that produces the largest decrease in impurity is chosen from a set of possible splits. Information gain is calculated by subtracting the weighted sum of the child node impurities from the impurity of the parent node. The lower the impurity of the child nodes, the bigger the information gain. Splitting continues until a maximum tree depth is reached (set by the *maxDepth* parameter), information gain greater than *minInfoGain* can no longer be achieved, or *minInstancesPerNode* is equal to the training instances produced per child nodes.

There are two impurity measures for classification (gini impurity and entropy) and one impurity measure for regression (variance). For classification, the default measure of impurity in Spark MLlib is gini impurity. The gini score is a metric that quantifies the purity of the node. A single class of data exists within a node if the gini score is equal to zero (the node is pure). If the gini score is greater than zero, it means that the node contains data that belong to different classes.

A decision tree is easy to interpret. In contrast with linear models like logistic regression, decision trees do not require feature scaling. It is able to handle missing features and works with both continuous and categorical features.[v] One-hot encoding categorical features[vi] are not required and are in fact discouraged when using decision trees and tree-based ensembles. One-hot encoding creates unbalanced trees and requires trees to grow extremely deep to achieve good predictive performance. This is especially true for high-cardinality categorical features.

On the downside, decision trees are sensitive to noise in the data and have a tendency to overfit. Due to this limitation, decision trees by themselves are rarely used in real-world production environments. Nowadays, decision trees serve as the base model for more powerful ensemble algorithms such as Random Forest and Gradient-Boosted Trees.

Random Forest

Random Forest is an ensemble algorithm that uses a collection of decision trees for classification and regression. It uses a method called *bagging* (or bootstrap aggregation) to reduce variance while maintaining low bias. Bagging trains individual trees from subsets of the training data. In addition to bagging, Random Forest uses another method called *feature bagging*. In contrast to bagging (using subsets of observations), feature bagging uses a subset of features (columns). Feature bagging aims to reduce the correlation between the decision trees. Without feature bagging, the individual trees will be extremely similar especially in situations where there are only a few dominant features.

For classification, a majority vote of the output, or the mode, of the individual trees becomes the final prediction of the model. For regression, the average of the output of the individual trees becomes the final output (Figure 3-3). Spark trains several trees in parallel since each tree is trained independently in Random Forest. I discuss Random Forest in more detail later in the chapter.

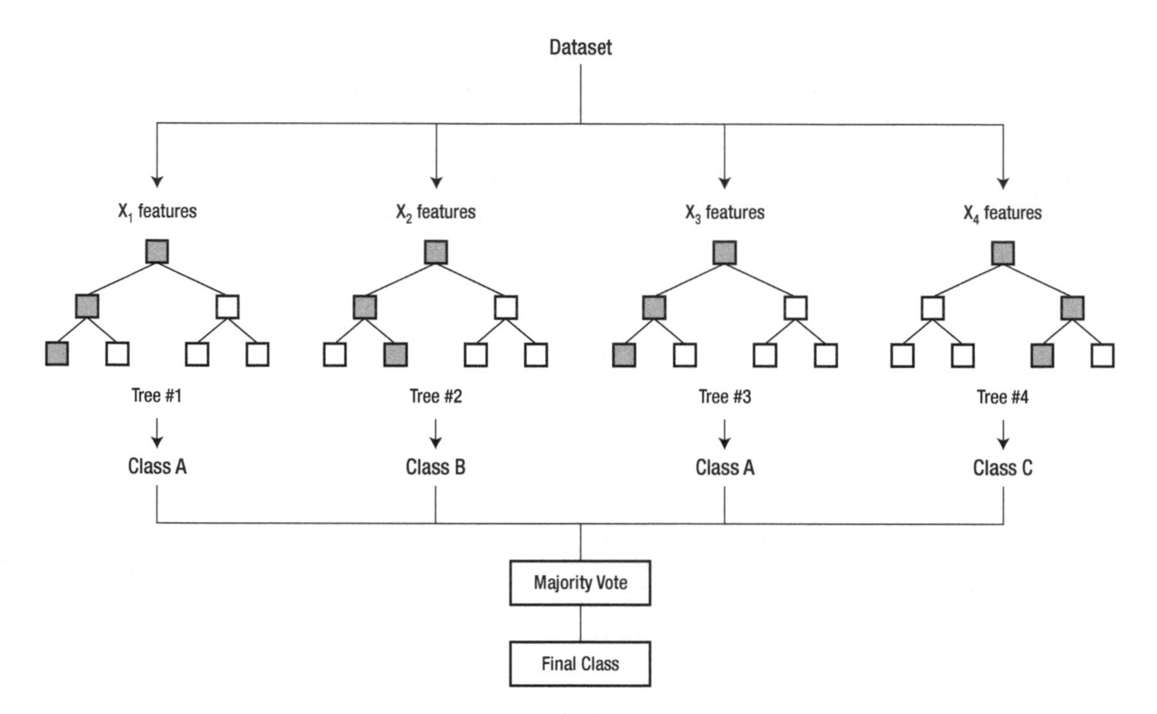

Figure 3-3. *Random Forest for classification*

Gradient-Boosted Trees

Gradient-Boosted Tree (GBT) is another tree-based ensemble algorithm similar to Random Forest. GBTs use a technique known as *boosting to* create a strong learner from weak learners (shallow trees). GBTs train an ensemble of decision trees sequentially[vii] with each succeeding tree decreasing the error of the previous tree. This is done by using the residuals of the previous model to fit the next model.[viii] This residual-correction process[ix] is performed a set number of iterations with the number of iterations determined by cross-validation, until the residuals have been fully minimized.

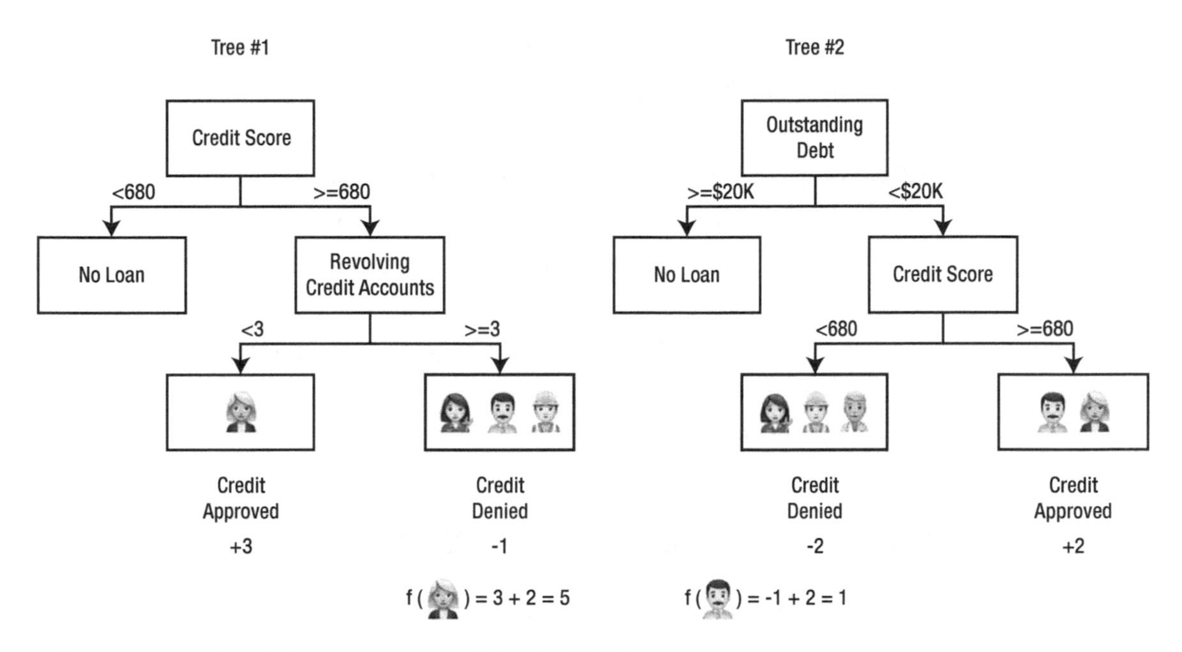

Figure 3-4. *Decision tree ensembles in GBTs*

Figure 3-4 shows how decision tree ensembles work in GBTs. Using our credit risk example, individuals are classified into different leaves based on their creditworthiness. Each leaf within the decision tree is assigned a score. The scores of multiple trees are added together to get the final prediction score. For instance, Figure 3-4 shows the first decision tree assigned the woman a score of 3. The second tree assigned her a score of 2. Adding both scores together gives the woman a final score of 5. Note that the decision trees complemented each other. This is one of the main principles of GBTs. Associating a score with each leaf provides GBTs with an integrated approach to optimization.[x]

Random Forests vs. Gradient-Boosted Trees

Since Gradient-Boosted Trees are trained sequentially, it is generally considered slower and less scalable than Random Forest, which is capable of training multiple trees in parallel. However, GBTs often use shallower trees compared to Random Forest which means GBTs can be trained faster.

Increasing the number of trees in GBTs increases the chance of overfitting (GBTs decrease bias by utilizing more trees), while increasing the number of trees in Random Forest decreases the chances of overfitting (Random Forests decrease variance by utilizing more trees). Generally speaking, adding more trees improves performance in

Random Forests, while GBTs' performance will start to degrade when the number of trees starts to grow too big.[xi] Because of this, GBTs can be harder to tune than Random Forest.

Gradient-Boosted Trees are generally considered more powerful than Random Forest if parameters are tuned correctly. GBTs add new decision trees that complement the previously constructed ones which results in better predictive accuracy with fewer trees compared to Random Forest.[xii]

Most of the newer algorithms developed in recent years for classification and regression such as XGBoost and LightGBM are improved variants of GBTs. They do not have the limitations of traditional GBTs.

Third-Party Classification and Regression Algorithms

Countless open source contributors have devoted time and effort in developing third-party machine learning algorithms for Spark. Although they are not part of the core Spark MLlib library, companies such as Databricks (XGBoost) and Microsoft (LightGBM) have put their support behind these projects and are used extensively around the world. XGBoost and LightGBM are currently considered the next-generation machine learning algorithms for classification and regression. They are the go-to algorithms in situations where accuracy and speed are critical. I will discuss both of them later in the chapter. For now, let's get our hands dirty and dive into some examples.

Multiclass Classification with Logistic Regression

Logistic regression is a linear classifier that predicts probabilities. It is popular for its ease of use and fast training speed and is frequently used for both binary classification and multiclass classification. A linear classifier such as logistic regression is suitable when your data has a clear decision boundary, as shown in the first chart of Figure 3-5. In cases where the classes are not linearly separable (as shown in the second chart), nonlinear classifiers such as tree-based ensembles should be considered.

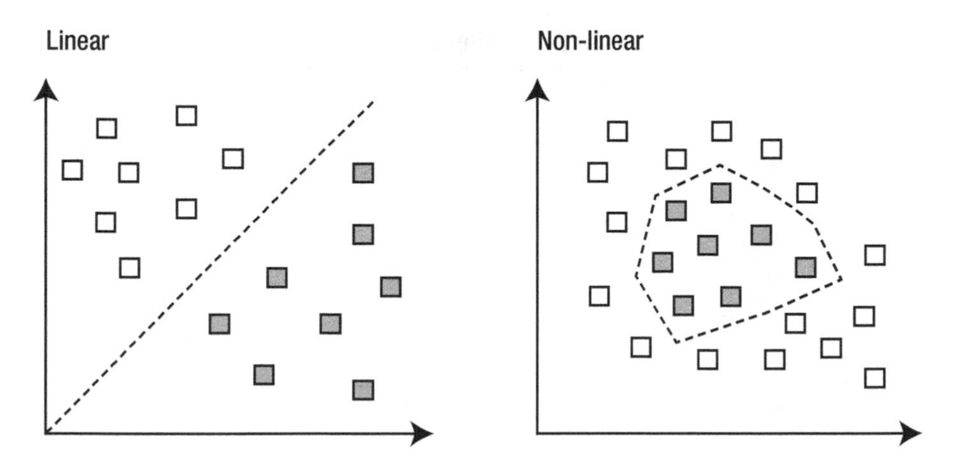

Figure 3-5. *Linear vs. nonlinear classification problems*

Example

We will work on a multiclass classification problem for our first example using the popular Iris dataset (see Listing 3-1). The dataset contains three classes of 50 instances each, where each class refers to a variety of iris plant (Iris Setosa, Iris Versicolor, and Iris Virginica). As you can see from Figure 3-6, Iris Setosa is linearly separable from Iris Versicolor and Iris Virginica, but Iris Versicolor and Iris Virginica are not linearly separable from each other. Logistic regression should still do a decent job at classifying the dataset.

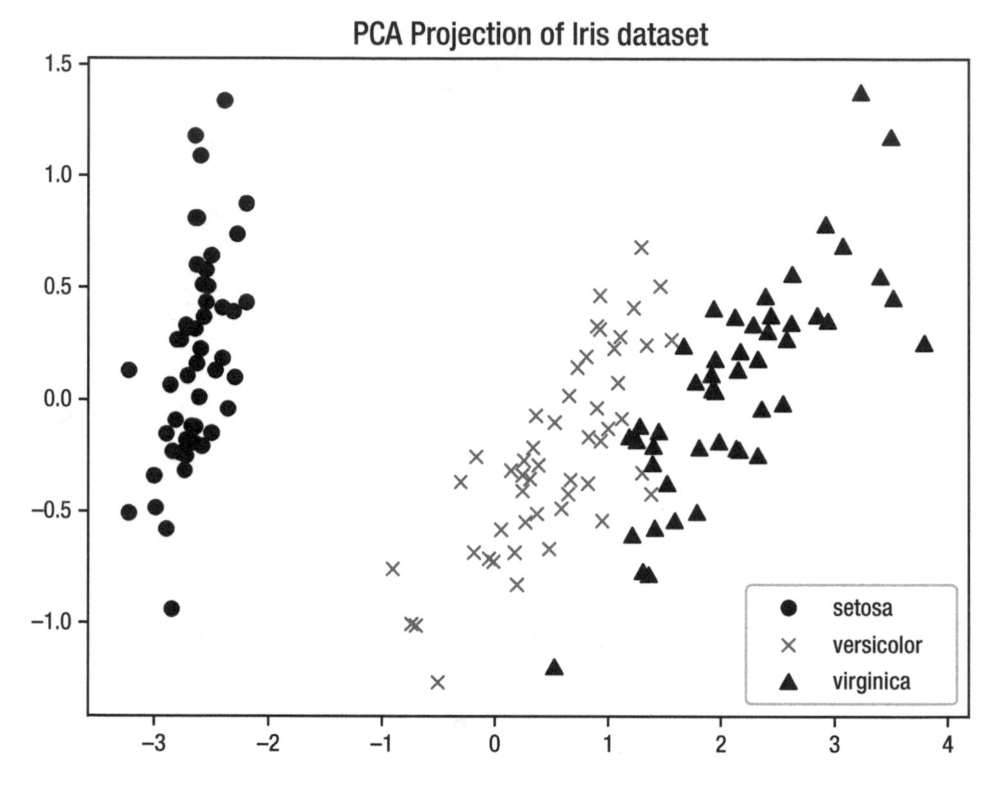

***Figure 3-6.** Principal component analysis projection of the Iris dataset*

Our goal is to predict the type of Iris plant given a set of features. The dataset contains four numeric features: sepal_length, sepal_width, petal_length, and petal_ width (all in centimeters).

Listing 3-1. Classification Using Logistic Regression

```
// Create a schema for our data.
import org.apache.spark.sql.types._

var irisSchema = StructType(Array (
    StructField("sepal_length",   DoubleType, true),
    StructField("sepal_width",   DoubleType, true),
    StructField("petal_length",   DoubleType, true),
    StructField("petal_width",   DoubleType, true),
    StructField("class",  StringType, true)
    ))
```

```
// Read the CSV file. Use the schema that we just defined.

val dataDF = spark.read.format("csv")
             .option("header","false")
             .schema(irisSchema)
             .load("/files/iris.data")

// Check the schema.

dataDF.printSchema

root
 |-- sepal_length: double (nullable = true)
 |-- sepal_width: double (nullable = true)
 |-- petal_length: double (nullable = true)
 |-- petal_width: double (nullable = true)
 |-- class: string (nullable = true)

// Inspect the data to make sure they're in the correct format.

dataDF.show

+------------+-----------+------------+-----------+-----------+
|sepal_length|sepal_width|petal_length|petal_width|      class|
+------------+-----------+------------+-----------+-----------+
|         5.1|        3.5|         1.4|        0.2|Iris-setosa|
|         4.9|        3.0|         1.4|        0.2|Iris-setosa|
|         4.7|        3.2|         1.3|        0.2|Iris-setosa|
|         4.6|        3.1|         1.5|        0.2|Iris-setosa|
|         5.0|        3.6|         1.4|        0.2|Iris-setosa|
|         5.4|        3.9|         1.7|        0.4|Iris-setosa|
|         4.6|        3.4|         1.4|        0.3|Iris-setosa|
|         5.0|        3.4|         1.5|        0.2|Iris-setosa|
|         4.4|        2.9|         1.4|        0.2|Iris-setosa|
|         4.9|        3.1|         1.5|        0.1|Iris-setosa|
|         5.4|        3.7|         1.5|        0.2|Iris-setosa|
|         4.8|        3.4|         1.6|        0.2|Iris-setosa|
|         4.8|        3.0|         1.4|        0.1|Iris-setosa|
|         4.3|        3.0|         1.1|        0.1|Iris-setosa|
```

```
|         5.8|         4.0|          1.2|         0.2|Iris-setosa|
|         5.7|         4.4|          1.5|         0.4|Iris-setosa|
|         5.4|         3.9|          1.3|         0.4|Iris-setosa|
|         5.1|         3.5|          1.4|         0.3|Iris-setosa|
|         5.7|         3.8|          1.7|         0.3|Iris-setosa|
|         5.1|         3.8|          1.5|         0.3|Iris-setosa|
+------------+-----------+------------+-----------+-----------+
only showing top 20 rows

// Calculate summary statistics for our data. This can
// be helpful in understanding the distribution of your data.

dataDF.describe().show(5,15)
+-------+---------------+---------------+---------------+---------------+
|summary|   sepal_length|    sepal_width|   petal_length|    petal_width|
+-------+---------------+---------------+---------------+---------------+
|  count|            150|            150|            150|            150|
|   mean|5.8433333333...|3.0540000000...|3.7586666666...|1.1986666666...|
| stddev|0.8280661279...|0.4335943113...|1.7644204199...|0.7631607417...|
|    min|            4.3|            2.0|            1.0|            0.1|
|    max|            7.9|            4.4|            6.9|            2.5|
+-------+---------------+---------------+---------------+---------------+

+--------------+
|         class|
+--------------+
|           150|
|          null|
|          null|
|   Iris-setosa|
|Iris-virginica|
+--------------+

// The input column class is currently a string. We'll use
// StringIndexer to encode it into a double. The new value
// will be stored in the new output column called label.
```

```
import org.apache.spark.ml.feature.StringIndexer

val labelIndexer = new StringIndexer()
                  .setInputCol("class")
                  .setOutputCol("label")

val dataDF2 = labelIndexer
             .fit(dataDF)
             .transform(dataDF)

// Check the schema of the new DataFrame.

dataDF2.printSchema
root
 |-- sepal_length: double (nullable = true)
 |-- sepal_width: double (nullable = true)
 |-- petal_length: double (nullable = true)
 |-- petal_width: double (nullable = true)
 |-- class: string (nullable = true)
 |-- label: double (nullable = false)

// Inspect the new column added to the DataFrame.

dataDF2.show

+------------+-----------+------------+-----------+-----------+-----+
|sepal_length|sepal_width|petal_length|petal_width|      class|label|
+------------+-----------+------------+-----------+-----------+-----+
|         5.1|        3.5|         1.4|        0.2|Iris-setosa|  0.0|
|         4.9|        3.0|         1.4|        0.2|Iris-setosa|  0.0|
|         4.7|        3.2|         1.3|        0.2|Iris-setosa|  0.0|
|         4.6|        3.1|         1.5|        0.2|Iris-setosa|  0.0|
|         5.0|        3.6|         1.4|        0.2|Iris-setosa|  0.0|
|         5.4|        3.9|         1.7|        0.4|Iris-setosa|  0.0|
|         4.6|        3.4|         1.4|        0.3|Iris-setosa|  0.0|
|         5.0|        3.4|         1.5|        0.2|Iris-setosa|  0.0|
|         4.4|        2.9|         1.4|        0.2|Iris-setosa|  0.0|
|         4.9|        3.1|         1.5|        0.1|Iris-setosa|  0.0|
|         5.4|        3.7|         1.5|        0.2|Iris-setosa|  0.0|
```

```
|         4.8|         3.4|         1.6|         0.2|Iris-setosa|  0.0|
|         4.8|         3.0|         1.4|         0.1|Iris-setosa|  0.0|
|         4.3|         3.0|         1.1|         0.1|Iris-setosa|  0.0|
|         5.8|         4.0|         1.2|         0.2|Iris-setosa|  0.0|
|         5.7|         4.4|         1.5|         0.4|Iris-setosa|  0.0|
|         5.4|         3.9|         1.3|         0.4|Iris-setosa|  0.0|
|         5.1|         3.5|         1.4|         0.3|Iris-setosa|  0.0|
|         5.7|         3.8|         1.7|         0.3|Iris-setosa|  0.0|
|         5.1|         3.8|         1.5|         0.3|Iris-setosa|  0.0|
+------------+-----------+------------+-----------+-----------+-----+
only showing top 20 rows

// Combine the features into a single vector
// column using the VectorAssembler transformer.

import org.apache.spark.ml.feature.VectorAssembler

val features = Array("sepal_length","sepal_width","petal_length",
"petal_width")


val assembler = new VectorAssembler()
                .setInputCols(features)
                .setOutputCol("features")

val dataDF3 = assembler.transform(dataDF2)

// Inspect the new column added to the DataFrame.

dataDF3.printSchema

root
 |-- sepal_length: double (nullable = true)
 |-- sepal_width: double (nullable = true)
 |-- petal_length: double (nullable = true)
 |-- petal_width: double (nullable = true)
 |-- class: string (nullable = true)
 |-- label: double (nullable = false)
 |-- features: vector (nullable = true)

// Inspect the new column added to the DataFrame.
```

```
dataDF3.show

+------------+-----------+------------+-----------+-----------+-----+
|sepal_length|sepal_width|petal_length|petal_width|      class|label|
+------------+-----------+------------+-----------+-----------+-----+
|         5.1|        3.5|         1.4|        0.2|Iris-setosa|  0.0|
|         4.9|        3.0|         1.4|        0.2|Iris-setosa|  0.0|
|         4.7|        3.2|         1.3|        0.2|Iris-setosa|  0.0|
|         4.6|        3.1|         1.5|        0.2|Iris-setosa|  0.0|
|         5.0|        3.6|         1.4|        0.2|Iris-setosa|  0.0|
|         5.4|        3.9|         1.7|        0.4|Iris-setosa|  0.0|
|         4.6|        3.4|         1.4|        0.3|Iris-setosa|  0.0|
|         5.0|        3.4|         1.5|        0.2|Iris-setosa|  0.0|
|         4.4|        2.9|         1.4|        0.2|Iris-setosa|  0.0|
|         4.9|        3.1|         1.5|        0.1|Iris-setosa|  0.0|
|         5.4|        3.7|         1.5|        0.2|Iris-setosa|  0.0|
|         4.8|        3.4|         1.6|        0.2|Iris-setosa|  0.0|
|         4.8|        3.0|         1.4|        0.1|Iris-setosa|  0.0|
|         4.3|        3.0|         1.1|        0.1|Iris-setosa|  0.0|
|         5.8|        4.0|         1.2|        0.2|Iris-setosa|  0.0|
|         5.7|        4.4|         1.5|        0.4|Iris-setosa|  0.0|
|         5.4|        3.9|         1.3|        0.4|Iris-setosa|  0.0|
|         5.1|        3.5|         1.4|        0.3|Iris-setosa|  0.0|
|         5.7|        3.8|         1.7|        0.3|Iris-setosa|  0.0|
|         5.1|        3.8|         1.5|        0.3|Iris-setosa|  0.0|
+------------+-----------+------------+-----------+-----------+-----+
+-----------------+
|         features|
+-----------------+
|[5.1,3.5,1.4,0.2]|
|[4.9,3.0,1.4,0.2]|
|[4.7,3.2,1.3,0.2]|
|[4.6,3.1,1.5,0.2]|
|[5.0,3.6,1.4,0.2]|
|[5.4,3.9,1.7,0.4]|
|[4.6,3.4,1.4,0.3]|
```

```
|[5.0,3.4,1.5,0.2]|
|[4.4,2.9,1.4,0.2]|
|[4.9,3.1,1.5,0.1]|
|[5.4,3.7,1.5,0.2]|
|[4.8,3.4,1.6,0.2]|
|[4.8,3.0,1.4,0.1]|
|[4.3,3.0,1.1,0.1]|
|[5.8,4.0,1.2,0.2]|
|[5.7,4.4,1.5,0.4]|
|[5.4,3.9,1.3,0.4]|
|[5.1,3.5,1.4,0.3]|
|[5.7,3.8,1.7,0.3]|
|[5.1,3.8,1.5,0.3]|
+-----------------+

only showing top 20 rows

// Let's measure the statistical dependence between
// the features and the class using Pearson correlation.
dataDF3.stat.corr("petal_length","label")
res48: Double = 0.9490425448523336

dataDF3.stat.corr("petal_width","label")
res49: Double = 0.9564638238016178

dataDF3.stat.corr("sepal_length","label")
res50: Double = 0.7825612318100821

dataDF3.stat.corr("sepal_width","label")
res51: Double = -0.41944620026002677

// The petal_length and petal_width have extremely high class correlation,
// while sepal_length and sepal_width have low class correlation.
// As discussed in Chapter 2, correlation evaluates how strong the linear
// relationship between two variables. You can use correlation to select
// relevant features (feature-class correlation) and identify redundant
// features (intra-feature correlation).
```

```
// Divide our dataset into training and test datasets.
val seed = 1234

val Array(trainingData, testData) = dataDF3.randomSplit(Array(0.8, 0.2), seed)

// We can now fit a model on the training dataset
// using logistic regression.

import org.apache.spark.ml.classification.LogisticRegression

val lr = new LogisticRegression()

// Train a model using our training dataset.

val model = lr.fit(trainingData)

// Predict on our test dataset.

val predictions = model.transform(testData)

// Note the new columns added to the DataFrame:
// rawPrediction, probability, prediction.

predictions.printSchema

root
 |-- sepal_length: double (nullable = true)
 |-- sepal_width: double (nullable = true)
 |-- petal_length: double (nullable = true)
 |-- petal_width: double (nullable = true)
 |-- class: string (nullable = true)
 |-- label: double (nullable = false)
 |-- features: vector (nullable = true)
 |-- rawPrediction: vector (nullable = true)
 |-- probability: vector (nullable = true)
 |-- prediction: double (nullable = false)

// Inspect the predictions.

predictions.select("sepal_length","sepal_width",
"petal_length","petal_width","label","prediction").show
```

```
+------------+-----------+------------+-----------+-----+----------+
|sepal_length|sepal_width|petal_length|petal_width|label|prediction|
+------------+-----------+------------+-----------+-----+----------+
|         4.3|        3.0|         1.1|        0.1|  0.0|       0.0|
|         4.4|        2.9|         1.4|        0.2|  0.0|       0.0|
|         4.4|        3.0|         1.3|        0.2|  0.0|       0.0|
|         4.8|        3.1|         1.6|        0.2|  0.0|       0.0|
|         5.0|        3.3|         1.4|        0.2|  0.0|       0.0|
|         5.0|        3.4|         1.5|        0.2|  0.0|       0.0|
|         5.0|        3.6|         1.4|        0.2|  0.0|       0.0|
|         5.1|        3.4|         1.5|        0.2|  0.0|       0.0|
|         5.2|        2.7|         3.9|        1.4|  1.0|       1.0|
|         5.2|        4.1|         1.5|        0.1|  0.0|       0.0|
|         5.3|        3.7|         1.5|        0.2|  0.0|       0.0|
|         5.6|        2.9|         3.6|        1.3|  1.0|       1.0|
|         5.8|        2.8|         5.1|        2.4|  2.0|       2.0|
|         6.0|        2.2|         4.0|        1.0|  1.0|       1.0|
|         6.0|        2.9|         4.5|        1.5|  1.0|       1.0|
|         6.0|        3.4|         4.5|        1.6|  1.0|       1.0|
|         6.2|        2.8|         4.8|        1.8|  2.0|       2.0|
|         6.2|        2.9|         4.3|        1.3|  1.0|       1.0|
|         6.3|        2.8|         5.1|        1.5|  2.0|       1.0|
|         6.7|        3.1|         5.6|        2.4|  2.0|       2.0|
+------------+-----------+------------+-----------+-----+----------+
only showing top 20 rows

// Inspect the rawPrediction and probability columns.

predictions.select("rawPrediction","probability","prediction")
          .show(false)
+----------------------------------------------------------+
|rawPrediction                                             |
+----------------------------------------------------------+
|[-27765.164694901094,17727.78535517628,10037.379339724806] |
|[-24491.649758932126,13931.526474094646,10560.123284837473]|
|[20141.806983153703,1877.784589255676,-22019.591572409383] |
```

```
|[-46255.06332259462,20994.503038678085,25260.560283916537]  |
|[25095.115980666546,110.99834659454791,-25206.114327261093] |
|[-41011.14350152455,17036.32945903473,23974.814042489823]   |
|[20524.55747106708,1750.139974552606,-22274.697445619684]   |
|[29601.783587714817,-1697.1845083924927,-27904.599079322325]|
|[38919.06696252647,-5453.963471106039,-33465.10349142042]   |
|[-39965.27448934488,17725.41646382807,22239.85802551682]    |
|[-18994.667253235268,12074.709651218403,6919.957602016859]  |
|[-43236.84898013162,18023.80837865029,25213.040601481334]   |
|[-31543.179893646557,16452.928101990834,15090.251791655724] |
|[-21666.087284218,13802.846783092147,7863.24050112584]      |
|[-24107.97243292983,14585.93668397567,9522.035748954155]    |
|[25629.52586174148,-192.40731255107312,-25437.11854919041]  |
|[-14271.522512385294,11041.861803401871,3229.660708983418]  |
|[-16548.06114507441,10139.917257827732,6408.143887246673]   |
|[22598.60355651257,938.4220993796007,-23537.025655892172]   |
|[-40984.78286289556,18297.704445848023,22687.078417047538]  |
+------------------------------------------------------------+

+-------------+----------+
|probability  |prediction|
+-------------+----------+
|[0.0,1.0,0.0]|1.0       |
|[0.0,1.0,0.0]|1.0       |
|[1.0,0.0,0.0]|0.0       |
|[0.0,0.0,1.0]|2.0       |
|[1.0,0.0,0.0]|0.0       |
|[0.0,0.0,1.0]|2.0       |
|[1.0,0.0,0.0]|0.0       |
|[1.0,0.0,0.0]|0.0       |
|[1.0,0.0,0.0]|0.0       |
|[0.0,0.0,1.0]|2.0       |
|[0.0,1.0,0.0]|1.0       |
|[0.0,1.0,0.0]|1.0       |
|[0.0,1.0,0.0]|1.0       |
|[1.0,0.0,0.0]|0.0       |
```

```
|[0.0,1.0,0.0]|1.0       |
|[0.0,1.0,0.0]|1.0       |
|[1.0,0.0,0.0]|0.0       |
|[0.0,0.0,1.0]|2.0       |
+-------------+----------+
only showing top 20 rows

// Evaluate the model. Several evaluation metrics are available
// for multiclass classification: f1 (default), accuracy,
// weightedPrecision, and weightedRecall.
// I discuss evaluation metrics in more detail in Chapter 2.

import org.apache.spark.ml.evaluation.MulticlassClassificationEvaluator

val evaluator = new MulticlassClassificationEvaluator().setMetricName("f1")

val f1 = evaluator.evaluate(predictions)

f1: Double = 0.958119658119658

val wp = evaluator.setMetricName("weightedPrecision").evaluate(predictions)

wp: Double = 0.9635416666666667

val wr = evaluator.setMetricName("weightedRecall").evaluate(predictions)

wr: Double = 0.9583333333333335

val accuracy = evaluator.setMetricName("accuracy").evaluate(predictions)

accuracy: Double = 0.9583333333333334
```

Logistic regression is a popular classification algorithm often used as a first baseline algorithm due to its speed and simplicity. For production use, more advanced tree-based ensembles is generally preferred due to their superior accuracy and ability to capture complex nonlinear relationships in datasets.

Churn Prediction with Random Forest

Random Forest is a powerful ensemble learning algorithm built on multiple decision trees as base models, each trained on different bootstrapped subsets of the data in parallel. As discussed earlier, decision trees tend to overfit. Random Forest addresses overfitting by using a technique called *bagging* (bootstrap aggregation) to train each decision tree with randomly selected subset of the data. Bagging decreases the variance of the model, helping avoid overfitting. Random Forest decreases the variance of the model without increasing the bias. It also performs *feature bagging*, which randomly selects features for each decision tree. The goal of feature bagging is to reduce the correlation between individual trees.

For classification, the final class is determined via majority voting. The mode (the most frequently occurring) of the classes produced by the individual decision trees becomes the final class. For regression, the average of the output of the individual decision trees becomes the final output of the model.

Since Random Forest uses decision trees as its base model, it inherits most of its qualities. It is capable of handling both continuous and categorical features and do not require feature scaling and one-hot encoding. Random Forest also performs well on imbalanced data since its hierarchical nature forces them to address both classes. Finally, Random Forest can capture nonlinear relationships between dependent and independent variables.

Random Forest is one of the most popular tree-based ensemble algorithms for classification and regression due to its interpretability, accuracy, and flexibility. However, training Random Forest models can be computationally intensive (which makes it ideal for parallelization on multicore and distributed environments such as Hadoop or Spark). It requires significantly more memory and computing resources compared to linear models such as logistic regression or naïve Bayes. Also, Random Forest tends to perform poorly on high-dimensional data such as text or genomic data.

Note The CSIRO Bioinformatics team developed a highly scalable implementation of Random Forest originally designed for high-dimensional genomic data known as VariantSpark RF.[xiii] VariantSpark RF can handle millions of features and was shown in benchmarks[xiv] to be significantly more scalable than MLlib's Random Forest implementation. More information about VariantSpark RF can be found on CSIRO's Bioinformatics web site. ReForeSt is another highly scalable implementation of Random Forest developed at SmartLab research laboratory at DIBRIS – University of Genoa, Italy.[xv] ReForeSt can handle millions of features and supports Random Forest Rotation, a new ensemble method that extends the classic Random Forest algorithm.[xvi]

Parameters

Random Forest is relatively easy to tune. Properly setting a few important parameters[xvii] is often enough to successfully use Random Forest.

- *max_depth:* Specifies the maximum depth of the tree. Setting a high value for max_depth can make the model more expressive, but setting it too high may increase the likelihood of overfitting and make the model more complex.
- *num_trees:* Specifies the numbers of trees to fit. Increasing the number of trees decreases variance, generally improving accuracy. Increasing the number trees can slow down training. Adding more trees beyond a certain point may not improve accuracy.
- *FeatureSubsetStrategy:* Specifies the fraction of features to use for splitting at each node. Setting this parameter can improve training speed.
- *subsamplingRate:* Specifies the fraction of data that will be selected for training each tree. Setting this parameter can improve training speed and help prevent overfitting. Setting it too low may cause underfitting.

I provided some general guidelines, but as always, performing a parameter grid search to determine the optimal values for these parameters is highly recommended. For a complete list of Random Forest parameters, consult Spark MLlib's online documentation.

Example

Churn prediction is an important classification use case for banks, insurance companies, telcos, cable TV operators, and streaming services such as Netflix, Hulu, Spotify, and Apple Music. Companies that can predict customers who are more likely to cancel subscription to their service can implement a more effective customer retention strategy. Keeping customers is valuable. Customer churn costs US businesses an estimated $136 billion per year according to a study done by a leading customer engagement analytics firm.[xviii] Research done by Bain & Company shows increasing customer retention rates by just 5% increases profits by 25% to 95%.[xix] Another statistic provided by Lee Resource Inc. says that attracting new customer will cost a company five times more than keeping an existing customer.[xx]

We'll use a popular telco churn dataset from the University of California Irvine Machine Learning Repository for our example (see Listing 3-2). This is a popular Kaggle dataset[xxi] and used extensively online.[xxii]

For most of the examples throughout the book, I will execute the transformers and estimators individually (instead of specifying them all in the pipeline) so you can see the new column added to the resulting DataFrame. This will help you see what's going on "under the hood" as you work through the examples.

Listing 3-2. Churn Prediction Using Random Forest

```
// Load the CSV file into a DataFrame.
val dataDF = spark.read.format("csv")
             .option("header", "true")
             .option("inferSchema", "true")
             .load("churn_data.txt")

// Check the schema.

dataDF.printSchema
root
 |-- state: string (nullable = true)
 |-- account_length: double (nullable = true)
 |-- area_code: double (nullable = true)
 |-- phone_number: string (nullable = true)
 |-- international_plan: string (nullable = true)
 |-- voice_mail_plan: string (nullable = true)
 |-- number_vmail_messages: double (nullable = true)
 |-- total_day_minutes: double (nullable = true)
 |-- total_day_calls: double (nullable = true)
 |-- total_day_charge: double (nullable = true)
 |-- total_eve_minutes: double (nullable = true)
 |-- total_eve_calls: double (nullable = true)
 |-- total_eve_charge: double (nullable = true)
 |-- total_night_minutes: double (nullable = true)
```

```
 |-- total_night_calls: double (nullable = true)
 |-- total_night_charge: double (nullable = true)
 |-- total_intl_minutes: double (nullable = true)
 |-- total_intl_calls: double (nullable = true)
 |-- total_intl_charge: double (nullable = true)
 |-- number_customer_service_calls: double (nullable = true)
 |-- churned: string (nullable = true)

// Select a few columns.

dataDF
.select("state","phone_number","international_plan","total_day_
minutes","churned").show
+-----+------------+------------------+-----------------+-------+
|state|phone_number|international_plan|total_day_minutes|churned|
+-----+------------+------------------+-----------------+-------+
|   KS|    382-4657|                no|            265.1|  False|
|   OH|    371-7191|                no|            161.6|  False|
|   NJ|    358-1921|                no|            243.4|  False|
|   OH|    375-9999|               yes|            299.4|  False|
|   OK|    330-6626|               yes|            166.7|  False|
|   AL|    391-8027|               yes|            223.4|  False|
|   MA|    355-9993|                no|            218.2|  False|
|   MO|    329-9001|               yes|            157.0|  False|
|   LA|    335-4719|                no|            184.5|  False|
|   WV|    330-8173|               yes|            258.6|  False|
|   IN|    329-6603|                no|            129.1|   True|
|   RI|    344-9403|                no|            187.7|  False|
|   IA|    363-1107|                no|            128.8|  False|
|   MT|    394-8006|                no|            156.6|  False|
|   IA|    366-9238|                no|            120.7|  False|
|   NY|    351-7269|                no|            332.9|   True|
|   ID|    350-8884|                no|            196.4|  False|
|   VT|    386-2923|                no|            190.7|  False|
|   VA|    356-2992|                no|            189.7|  False|
|   TX|    373-2782|                no|            224.4|  False|
+-----+------------+------------------+-----------------+-------+
```

```
only showing top 20 rows

import org.apache.spark.ml.feature.StringIndexer

// Convert the string column "churned" ("True", "False") to double (1,0).

val labelIndexer = new StringIndexer()
                   .setInputCol("churned")
                   .setOutputCol("label")

// Convert the string column "international_plan" ("yes", "no")
// to double 1,0.

val intPlanIndexer = new StringIndexer()
                     .setInputCol("international_plan")
                     .setOutputCol("int_plan")

// Let's select our features. Domain knowledge is essential in feature
// selection. I would think total_day_minutes and total_day_calls have
// some influence on customer churn. A significant drop in these two
// metrics might indicate that the customer does not need the service
// any longer and may be on the verge of cancelling their phone plan.
// However, I don't think phone_number, area_code, and state have any
// predictive qualities at all. We discuss feature selection later in
// this chapter.

val features = Array("number_customer_service_calls","total_day_
minutes","total_eve_minutes","account_length","number_vmail_
messages","total_day_calls","total_day_charge","total_eve_calls","total_
eve_charge","total_night_calls","total_intl_calls","total_intl_
charge","int_plan")

// Combine a given list of columns into a single vector column
// including all the features needed to train ML models.

import org.apache.spark.ml.feature.VectorAssembler

val assembler = new VectorAssembler()
                .setInputCols(features)
                .setOutputCol("features")
```

```
// Add the label column to the DataFrame.

val dataDF2 = labelIndexer
              .fit(dataDF)
              .transform(dataDF)

dataDF2.printSchema

root
 |-- state: string (nullable = true)
 |-- account_length: double (nullable = true)
 |-- area_code: double (nullable = true)
 |-- phone_number: string (nullable = true)
 |-- international_plan: string (nullable = true)
 |-- voice_mail_plan: string (nullable = true)
 |-- number_vmail_messages: double (nullable = true)
 |-- total_day_minutes: double (nullable = true)
 |-- total_day_calls: double (nullable = true)
 |-- total_day_charge: double (nullable = true)
 |-- total_eve_minutes: double (nullable = true)
 |-- total_eve_calls: double (nullable = true)
 |-- total_eve_charge: double (nullable = true)
 |-- total_night_minutes: double (nullable = true)
 |-- total_night_calls: double (nullable = true)
 |-- total_night_charge: double (nullable = true)
 |-- total_intl_minutes: double (nullable = true)
 |-- total_intl_calls: double (nullable = true)
 |-- total_intl_charge: double (nullable = true)
 |-- number_customer_service_calls: double (nullable = true)
 |-- churned: string (nullable = true)
 |-- label: double (nullable = false)

// "True" was converted to 1 and "False" was converted to 0.

dataDF2.select("churned","label").show
```

```
+-------+-----+
|churned|label|
+-------+-----+
|  False|  0.0|
|  False|  0.0|
|  False|  0.0|
|  False|  0.0|
|  False|  0.0|
|  False|  0.0|
|  False|  0.0|
|  False|  0.0|
|  False|  0.0|
|  False|  0.0|
|   True|  1.0|
|  False|  0.0|
|  False|  0.0|
|  False|  0.0|
|  False|  0.0|
|   True|  1.0|
|  False|  0.0|
|  False|  0.0|
|  False|  0.0|
|  False|  0.0|
+-------+-----+
only showing top 20 rows

// Add the int_plan column to the DataFrame.

val dataDF3 = intPlanIndexer.fit(dataDF2).transform(dataDF2)

dataDF3.printSchema

root
 |-- state: string (nullable = true)
 |-- account_length: double (nullable = true)
 |-- area_code: double (nullable = true)
 |-- phone_number: string (nullable = true)
 |-- international_plan: string (nullable = true)
```

```
 |-- voice_mail_plan: string (nullable = true)
 |-- number_vmail_messages: double (nullable = true)
 |-- total_day_minutes: double (nullable = true)
 |-- total_day_calls: double (nullable = true)
 |-- total_day_charge: double (nullable = true)
 |-- total_eve_minutes: double (nullable = true)
 |-- total_eve_calls: double (nullable = true)
 |-- total_eve_charge: double (nullable = true)
 |-- total_night_minutes: double (nullable = true)
 |-- total_night_calls: double (nullable = true)
 |-- total_night_charge: double (nullable = true)
 |-- total_intl_minutes: double (nullable = true)
 |-- total_intl_calls: double (nullable = true)
 |-- total_intl_charge: double (nullable = true)
 |-- number_customer_service_calls: double (nullable = true)
 |-- churned: string (nullable = true)
 |-- label: double (nullable = false)
 |-- int_plan: double (nullable = false)

dataDF3.select("international_plan","int_plan").show

+------------------+--------+
|international_plan|int_plan|
+------------------+--------+
|                no|     0.0|
|                no|     0.0|
|                no|     0.0|
|               yes|     1.0|
|               yes|     1.0|
|               yes|     1.0|
|                no|     0.0|
|               yes|     1.0|
|                no|     0.0|
|               yes|     1.0|
|                no|     0.0|
|                no|     0.0|
|                no|     0.0|
```

```
|                no|     0.0|
|                no|     0.0|
|                no|     0.0|
|                no|     0.0|
|                no|     0.0|
|                no|     0.0|
|                no|     0.0|
+------------------+--------+
only showing top 20 rows

// Add the features vector column to the DataFrame.

val dataDF4 = assembler.transform(dataDF3)

dataDF4.printSchema

root
 |-- state: string (nullable = true)
 |-- account_length: double (nullable = true)
 |-- area_code: double (nullable = true)
 |-- phone_number: string (nullable = true)
 |-- international_plan: string (nullable = true)
 |-- voice_mail_plan: string (nullable = true)
 |-- number_vmail_messages: double (nullable = true)
 |-- total_day_minutes: double (nullable = true)
 |-- total_day_calls: double (nullable = true)
 |-- total_day_charge: double (nullable = true)
 |-- total_eve_minutes: double (nullable = true)
 |-- total_eve_calls: double (nullable = true)
 |-- total_eve_charge: double (nullable = true)
 |-- total_night_minutes: double (nullable = true)
 |-- total_night_calls: double (nullable = true)
 |-- total_night_charge: double (nullable = true)
 |-- total_intl_minutes: double (nullable = true)
 |-- total_intl_calls: double (nullable = true)
 |-- total_intl_charge: double (nullable = true)
 |-- number_customer_service_calls: double (nullable = true)
 |-- churned: string (nullable = true)
```

```
 |-- label: double (nullable = false)
 |-- int_plan: double (nullable = false)
 |-- features: vector (nullable = true)

// The features have been vectorized.

dataDF4.select("features").show(false)

+----------------------------------------------------------------------+
|features                                                              |
+----------------------------------------------------------------------+
|[1.0,265.1,197.4,128.0,25.0,110.0,45.07,99.0,16.78,91.0,3.0,2.7,0.0]  |
|[1.0,161.6,195.5,107.0,26.0,123.0,27.47,103.0,16.62,103.0,3.0,3.7,0.0]|
|[0.0,243.4,121.2,137.0,0.0,114.0,41.38,110.0,10.3,104.0,5.0,3.29,0.0] |
|[2.0,299.4,61.9,84.0,0.0,71.0,50.9,88.0,5.26,89.0,7.0,1.78,1.0]       |
|[3.0,166.7,148.3,75.0,0.0,113.0,28.34,122.0,12.61,121.0,3.0,2.73,1.0] |
|[0.0,223.4,220.6,118.0,0.0,98.0,37.98,101.0,18.75,118.0,6.0,1.7,1.0]  |
|[3.0,218.2,348.5,121.0,24.0,88.0,37.09,108.0,29.62,118.0,7.0,2.03,0.0]|
|[0.0,157.0,103.1,147.0,0.0,79.0,26.69,94.0,8.76,96.0,6.0,1.92,1.0]    |
|[1.0,184.5,351.6,117.0,0.0,97.0,31.37,80.0,29.89,90.0,4.0,2.35,0.0]   |
|[0.0,258.6,222.0,141.0,37.0,84.0,43.96,111.0,18.87,97.0,5.0,3.02,1.0] |
|[4.0,129.1,228.5,65.0,0.0,137.0,21.95,83.0,19.42,111.0,6.0,3.43,0.0]  |
|[0.0,187.7,163.4,74.0,0.0,127.0,31.91,148.0,13.89,94.0,5.0,2.46,0.0]  |
|[1.0,128.8,104.9,168.0,0.0,96.0,21.9,71.0,8.92,128.0,2.0,3.02,0.0]    |
|[3.0,156.6,247.6,95.0,0.0,88.0,26.62,75.0,21.05,115.0,5.0,3.32,0.0]   |
|[4.0,120.7,307.2,62.0,0.0,70.0,20.52,76.0,26.11,99.0,6.0,3.54,0.0]    |
|[4.0,332.9,317.8,161.0,0.0,67.0,56.59,97.0,27.01,128.0,9.0,1.46,0.0]  |
|[1.0,196.4,280.9,85.0,27.0,139.0,33.39,90.0,23.88,75.0,4.0,3.73,0.0]  |
|[3.0,190.7,218.2,93.0,0.0,114.0,32.42,111.0,18.55,121.0,3.0,2.19,0.0] |
|[1.0,189.7,212.8,76.0,33.0,66.0,32.25,65.0,18.09,108.0,5.0,2.7,0.0]   |
|[1.0,224.4,159.5,73.0,0.0,90.0,38.15,88.0,13.56,74.0,2.0,3.51,0.0]    |
+----------------------------------------------------------------------+
only showing top 20 rows

// Split the data into training and test data.

val seed = 1234

val Array(trainingData, testData) = dataDF4.randomSplit(Array(0.8, 0.2), seed)
```

```
trainingData.count
res13: Long = 4009

testData.count
res14: Long = 991

// Create a Random Forest classifier.

import org.apache.spark.ml.classification.RandomForestClassifier

val rf = new RandomForestClassifier()
        .setFeatureSubsetStrategy("auto")
        .setSeed(seed)

// Create a binary classification evaluator, and set label column to
// be used for evaluation.

import org.apache.spark.ml.evaluation.BinaryClassificationEvaluator

val evaluator = new BinaryClassificationEvaluator().setLabelCol("label")

// Create a parameter grid.

import org.apache.spark.ml.tuning.ParamGridBuilder

val paramGrid = new ParamGridBuilder()
                .addGrid(rf.maxBins, Array(10, 20,30))
                .addGrid(rf.maxDepth, Array(5, 10, 15))
                .addGrid(rf.numTrees, Array(3, 5, 100))
                .addGrid(rf.impurity, Array("gini", "entropy"))
                .build()

// Create a pipeline.
import org.apache.spark.ml.Pipeline

val pipeline = new Pipeline().setStages(Array(rf))

// Create a cross-validator.

import org.apache.spark.ml.tuning.CrossValidator

val cv = new CrossValidator()
         .setEstimator(pipeline)
         .setEvaluator(evaluator)
```

```
        .setEstimatorParamMaps(paramGrid)
        .setNumFolds(3)

// We can now fit the model using the training dataset, choosing the
// best set of parameters for the model.

val model = cv.fit(trainingData)

// You can now make some predictions on our test data.

val predictions = model.transform(testData)

// Evaluate the model.

import org.apache.spark.ml.param.ParamMap

val pmap = ParamMap(evaluator.metricName -> "areaUnderROC")

val auc = evaluator.evaluate(predictions, pmap)

auc: Double = 0.9270599683335483

// Our Random Forest classifier has a high AUC score. The test
// data consists of 991 observations. 92 customers are predicted
// to leave the service.

predictions.count
res25: Long = 991

predictions.filter("prediction=1").count
res26: Long = 92

println(s"True Negative: ${predictions.select("*").where("prediction = 0
AND label = 0").count()}  True Positive: ${predictions.select("*").
where("prediction = 1 AND label = 1").count()}")

True Negative: 837 True Positive: 81

// Our test predicted 81 customers leaving who actually did leave and also
// predicted 837 customers not leaving who actually did not leave.

println(s"False Negative: ${predictions.select("*").where("prediction = 0
AND label = 1").count()} False Positive: ${predictions.select("*").
where("prediction = 1 AND label = 0").count()}")

False Negative: 62 False Positive: 11
```

```
// Our test predicted 11 customers leaving who actually did not leave and
// also predicted 62 customers not leaving who actually did leave.

// You can sort the output by RawPrediction or Probability to target
// highest-probability customers. RawPrediction and Probability
// provide a measure of confidence for each prediction. The larger
// the value, the more confident the model is in its prediction.

predictions.select("phone_number","RawPrediction","prediction")
           .orderBy($"RawPrediction".asc)
           .show(false)

+------------+--------------------------------------+----------+
|phone_number|RawPrediction                         |prediction|
+------------+--------------------------------------+----------+
| 366-1084   |[15.038138063913935,84.96186193608602]|1.0       |
| 334-6519   |[15.072688486480072,84.9273115135199] |1.0       |
| 359-5574   |[15.276260309388752,84.72373969061123]|1.0       |
| 399-7865   |[15.429722388653014,84.57027761134698]|1.0       |
| 335-2967   |[16.465107279664032,83.53489272033593]|1.0       |
| 345-9140   |[16.53288465159445,83.46711534840551] |1.0       |
| 342-6864   |[16.694165016887318,83.30583498311265]|1.0       |
| 419-1863   |[17.594670105674677,82.4053298943253] |1.0       |
| 384-7176   |[17.92764148018115,82.07235851981882] |1.0       |
| 357-1938   |[18.8550074623437,81.1449925376563]   |1.0       |
| 355-6837   |[19.556608109022648,80.44339189097732]|1.0       |
| 417-1488   |[20.13305147603522,79.86694852396475] |1.0       |
| 394-5489   |[21.05074084178182,78.94925915821818] |1.0       |
| 394-7447   |[21.376663858426735,78.62333614157326]|1.0       |
| 339-6477   |[21.549262081786424,78.45073791821355]|1.0       |
| 406-7844   |[21.92209788389343,78.07790211610656] |1.0       |
| 372-4073   |[22.098599119168263,77.90140088083176]|1.0       |
| 404-4809   |[22.515513847987147,77.48448615201283]|1.0       |
| 347-8659   |[22.66840460762997,77.33159539237005] |1.0       |
| 335-1874   |[23.336632598761128,76.66336740123884]|1.0       |
+------------+--------------------------------------+----------+
only showing top 20 rows
```

Feature Importance

Random Forest (and other tree-based ensembles) has built-in feature selection capability that can be used to measure the importance of each feature in the dataset (see Listing 3-3).

Random Forest calculates feature importance as the sum of the decrease in node impurity of each node across every tree aggregated every time a feature is chosen to split a node, divided by the number of trees in the forest. Spark MLlib provides a method that returns an estimate of the importance of each feature.

Listing 3-3. Showing Feature Importance with Random Forest

```
import org.apache.spark.ml.classification.RandomForestClassificationModel
import org.apache.spark.ml.PipelineModel

val bestModel = model.bestModel

val model = bestModel
            .asInstanceOf[PipelineModel]
            .stages
            .last
            .asInstanceOf[RandomForestClassificationModel]

model.featureImportances

feature_importances: org.apache.spark.ml.linalg.Vector =
(13,[0,1,2,3,4,5,6,7,8,9,10,11,12],
[
0.20827010117447803,
0.1667170878866465,
0.06099491253318444,
0.008184141410796346,
0.06664053647245761,
0.0072108752126555,
0.21097011684691344,
0.006902059667276019,
0.06831916361401609,
```

```
0.00644772968425685,
0.04105403721675372,
0.056954219262186724,
0.09133501901837866])
```

We get a vector in return containing the number of features (13 in our example), the array index of our features, and the corresponding weight. Table 3-1 shows the output in a more readable format with the actual features shown with the corresponding weight. As you can see, total_day_charge, total_day_minutes, and number_customer_service_calls are the most important features. It makes sense. A high number of customer service calls may indicate multiple disruptions to the service or a high number of customer complaints. Low total_day_minutes and total_day_charge may indicate that the customer is not using his phone plan that much, which could mean that he's getting ready to cancel his plan soon.

Table 3-1. *Feature Importance for Our Telco Churn Prediction Example*

Index	Feature	Feature Importance
0	number_customer_service_calls	0.20827010117447803
1	total_day_minutes	0.1667170878866465
2	total_eve_minutes	0.06099491253318444
3	account_length	0.008184141410796346
4	number_vmail_messages	0.06664053647245761
5	total_day_calls	0.0072108752126555
6	total_day_charge	0.21097011684691344
7	total_eve_calls	0.006902059667276019
8	total_eve_charge	0.06831916361401609
9	total_night_calls	0.00644772968425685
10	total_intl_calls	0.04105403721675372
11	total_intl_charge	0.056954219262186724
12	int_plan	0.09133501901837866

Note Spark MLlib's implementation of feature importance in Random Forest is also known as gini-based importance or mean decrease in impurity (MDI). Some implementation of Random Forest utilizes a different method to calculate feature importance known as accuracy-based importance or mean decrease in accuracy (MDA).[xxiii] Accuracy-based importance is calculated based on decreases in prediction accuracy as features are randomly permuted. Although Spark MLlib's implementation of Random Forest doesn't directly support this method, it is fairly straightforward to implement manually by evaluating the model while permuting the values of each feature one column at a time.

It is sometimes useful to examine the parameters used by the best model (see Listing 3-4).

Listing 3-4. Extracting the Parameters of the Random Forest Model

```
import org.apache.spark.ml.classification.RandomForestClassificationModel
import org.apache.spark.ml.PipelineModel

val bestModel = model
                .bestModel
                .asInstanceOf[PipelineModel]
                .stages
                .last
                .asInstanceOf[RandomForestClassificationModel]

 print(bestModel.extractParamMap)
{
        rfc_81c4d3786152-cacheNodeIds: false,
        rfc_81c4d3786152-checkpointInterval: 10,
        rfc_81c4d3786152-featureSubsetStrategy: auto,
        rfc_81c4d3786152-featuresCol: features,
        rfc_81c4d3786152-impurity: gini,
        rfc_81c4d3786152-labelCol: label,
        rfc_81c4d3786152-maxBins: 10,
        rfc_81c4d3786152-maxDepth: 15,
        rfc_81c4d3786152-maxMemoryInMB: 256,
```

```
        rfc_81c4d3786152-minInfoGain: 0.0,
        rfc_81c4d3786152-minInstancesPerNode: 1,
        rfc_81c4d3786152-numTrees: 100,
        rfc_81c4d3786152-predictionCol: prediction,
        rfc_81c4d3786152-probabilityCol: probability,
        rfc_81c4d3786152-rawPredictionCol: rawPrediction,
        rfc_81c4d3786152-seed: 1234,
        rfc_81c4d3786152-subsamplingRate: 1.0
}
```

eXtreme Gradient Boosting with XGBoost4J-Spark

Gradient boosting algorithms are some of the most powerful machine learning algorithms for classification and regression. There are currently various implementations of gradient boosting algorithms. Popular implementations include AdaBoost and CatBoost (a recently open sourced gradient boosting library from Yandex). Spark MLlib also includes its own gradient-boosted tree (GBT) implementation.

XGBoost (eXtreme Gradient Boosting) is one of the best gradient-boosted tree implementations currently available. Released on March 27, 2014, by Tianqi Chen as a research project, XGBoost has become the dominant machine learning algorithm for classification and regression. Designed for efficiency and scalability, its parallel tree boosting capabilities make it significantly faster than other tree-based ensemble algorithms. Due to its high accuracy, XGBoost has gained popularity by winning several machine learning competitions. In 2015, 17 out of the 29 winning solutions on Kaggle used XGBoost. All the top 10 solutions at the KDD Cup 2015[xxiv] used XGBoost.

XGBoost was designed using the general principles of gradient boosting, combining weak learners into a strong learner. But while gradient-boosted trees are built sequentially – slowly learning from data to improve its prediction in succeeding iteration, XGBoost builds trees in parallel. XGBoost produces better prediction performance by controlling model complexity and reducing overfitting through its built-in regularization. It uses an approximate algorithm to find split points when finding the best split points for a continuous feature.[xxv]

The approximate splitting method uses discrete bins to bucket continuous features, significantly speeding up model training. XGBoost includes another tree growing method using a histogram-based algorithm which provides an even more efficient method of bucketing continuous features into discrete bins. But while the approximate method creates a new set of bins per iteration, the histogram-based approach reuses bins over multiple iterations.

This approach allows for additional optimizations that are not achievable with the approximate method, such as the ability to cache bins and parent and sibling histogram subtraction.[xxvi] To optimize sorting operations, XGBoost stores sorted data in in-memory units of blocks. Sorting blocks can be efficiently distributed and performed by parallel CPU cores. XGBoost can effectively handle weighted data via its weighted quantile sketch algorithm, can efficiently handle sparse data, is cache aware, and supports out-of-core computing by utilizing disk space for large datasets so data does not have to fit in memory.

The XGBoost4J-Spark project was started in late 2016 to port XGBoost to Spark. XGBoost4J-Spark takes advantage of Spark's highly scalable distributed processing engine and is fully compatible with Spark MLlib's DataFrame/Dataset abstraction. XGBoost4J-Spark can be seamlessly embedded in a Spark MLlib pipeline and integrated with Spark MLlib's transformers and estimators.

Note XGBoost4J-Spark requires Apache Spark 2.4+. It is recommended to install Spark directly from `http://spark.apache.org`. XGBoost4J-Spark is not guaranteed to work with third-party Spark distributions from other vendors such as Cloudera, Hortonworks, or MapR. Consult your vendor's documentation for more information.[xxvii]

Parameters

XGBoost has a lot more parameters than Random Forest and generally requires more tuning. Initially focusing on the most important parameters can get you started with XGBoost. You can learn the rest as you become more accustomed with the algorithm.

- *max_depth:* Specifies the maximum depth of the tree. Setting a high value for max_depth may increase the likelihood of overfitting and make the model more complex.
- *n_estimators:* Specifies the numbers of trees to fit. Generally speaking, the larger value, the better. Setting this parameter too high may affect training speed. Adding more trees beyond a certain point may not improve accuracy. The default is set to 100.[xxviii]

- *sub_sample:* Specifies the fraction of data that will be selected for each tree. Setting this parameter can increase training speed and help prevent overfitting. Setting it too low may cause underfitting.
- *colsample_bytree:* Specifies the fraction of columns that will be selected randomly for each tree. Setting this parameter can increase training speed and help prevent overfitting. Related parameters include colsample_bylevel and colsample_bynode.
- *objective:* Specifies the learning task and learning objective. It is important to set the correct value for this parameter to avoid unpredictable results or poor accuracy. XGBClassifier defaults to *binary:logistic* for binary classification, while XGBRegressor defaults to *reg:squarederror*. Other values include *multi:softmax* and *multi:softprob* for multiclass classification; *rank:pairwise, rank:ndcg, and rank:map* for ranking; and *survival:cox* for survival regression using Cox proportional hazards model, to mention a few.
- *learning_rate (eta):* learning_rate is used as a shrinkage factor to reduce the feature weights after each boosting step, with the goal of slowing down the learning rate. This parameter is used to control overfitting. Lower values require more trees.
- *n_jobs:* Specifies the number of parallel threads used by XGBoost (if n_thread is deprecated, use this parameter instead).

These are just general guidelines on how to use the parameters. Performing a parameter grid search to determine the optimal values for these parameters is highly recommended. For a complete list of XGBoost parameters, consult XGBoost's online documentation.

Note To stay consistent with Scala's variable naming convention, XGBoost4J-Spark supports both the default set of parameters and the camel case variations of these parameters (e.g., max_depth and maxDepth).

Example

We will reuse the same telco churn dataset and most of the code from the previous Random Forest example (see Listing 3-5). This time, we will use a pipeline to tie the transformers and estimators together.

Listing 3-5. Churn Prediction Using XGBoost4J-Spark

```
// XGBoost4J-Spark is available as an external package.
// Start spark-shell. Specify the XGBoost4J-Spark package.

spark-shell --packages ml.dmlc:xgboost4j-spark:0.81

// Load the CSV file into a DataFrame.

val dataDF = spark.read.format("csv")
             .option("header", "true")
             .option("inferSchema", "true")
             .load("churn_data.txt")

// Check the schema.

dataDF.printSchema

root
 |-- state: string (nullable = true)
 |-- account_length: double (nullable = true)
 |-- area_code: double (nullable = true)
 |-- phone_number: string (nullable = true)
 |-- international_plan: string (nullable = true)
 |-- voice_mail_plan: string (nullable = true)
 |-- number_vmail_messages: double (nullable = true)
 |-- total_day_minutes: double (nullable = true)
 |-- total_day_calls: double (nullable = true)
 |-- total_day_charge: double (nullable = true)
 |-- total_eve_minutes: double (nullable = true)
 |-- total_eve_calls: double (nullable = true)
 |-- total_eve_charge: double (nullable = true)
 |-- total_night_minutes: double (nullable = true)
```

```
 |-- total_night_calls: double (nullable = true)
 |-- total_night_charge: double (nullable = true)
 |-- total_intl_minutes: double (nullable = true)
 |-- total_intl_calls: double (nullable = true)
 |-- total_intl_charge: double (nullable = true)
 |-- number_customer_service_calls: double (nullable = true)
 |-- churned: string (nullable = true)

// Select a few columns.

dataDF.select("state","phone_number","international_plan","churned").show

+-----+------------+------------------+-------+
|state|phone_number|international_plan|churned|
+-----+------------+------------------+-------+
|   KS|    382-4657|                no|  False|
|   OH|    371-7191|                no|  False|
|   NJ|    358-1921|                no|  False|
|   OH|    375-9999|               yes|  False|
|   OK|    330-6626|               yes|  False|
|   AL|    391-8027|               yes|  False|
|   MA|    355-9993|                no|  False|
|   MO|    329-9001|               yes|  False|
|   LA|    335-4719|                no|  False|
|   WV|    330-8173|               yes|  False|
|   IN|    329-6603|                no|   True|
|   RI|    344-9403|                no|  False|
|   IA|    363-1107|                no|  False|
|   MT|    394-8006|                no|  False|
|   IA|    366-9238|                no|  False|
|   NY|    351-7269|                no|   True|
|   ID|    350-8884|                no|  False|
|   VT|    386-2923|                no|  False|
|   VA|    356-2992|                no|  False|
|   TX|    373-2782|                no|  False|
+-----+------------+------------------+-------+
only showing top 20 rows
```

```
import org.apache.spark.ml.feature.StringIndexer

// Convert the String "churned" column ("True", "False") to double(1,0).

val labelIndexer = new StringIndexer()
                   .setInputCol("churned")
                   .setOutputCol("label")

// Convert the String "international_plan" ("no", "yes") column to
double(1,0).

val intPlanIndexer = new StringIndexer()
                     .setInputCol("international_plan")
                     .setOutputCol("int_plan")

// Specify features to be selected for model fitting.

val features = Array("number_customer_service_calls","total_day_
minutes","total_eve_minutes","account_length","number_vmail_
messages","total_day_calls","total_day_charge","total_eve_calls","total_
eve_charge","total_night_calls","total_intl_calls","total_intl_
charge","int_plan")

// Combines the features into a single vector column.

import org.apache.spark.ml.feature.VectorAssembler

val assembler = new VectorAssembler()
                .setInputCols(features)
                .setOutputCol("features")

// Split the data into training and test data.

val seed = 1234

val Array(trainingData, testData) = dataDF.randomSplit(Array(0.8, 0.2), seed)

// Create an XGBoost classifier.

import ml.dmlc.xgboost4j.scala.spark.XGBoostClassifier
import ml.dmlc.xgboost4j.scala.spark.XGBoostClassificationModel
```

```
val xgb = new XGBoostClassifier()
          .setFeaturesCol("features")
          .setLabelCol("label")

// XGBClassifier's objective parameter defaults to binary:logistic which
// is the learning task and objective that we want for this example
// (binary classification). Depending on your task, remember to set the
// correct learning task and objective.

import org.apache.spark.ml.evaluation.MulticlassClassificationEvaluator

val evaluator = new MulticlassClassificationEvaluator().
setLabelCol("label")

import org.apache.spark.ml.tuning.ParamGridBuilder

val paramGrid = new ParamGridBuilder()
                .addGrid(xgb.maxDepth, Array(3, 8))
                .addGrid(xgb.eta, Array(0.2, 0.6))
                .build()

// This time we'll specify all the steps in the pipeline.

import org.apache.spark.ml.{ Pipeline, PipelineStage }

val pipeline = new Pipeline()
               .setStages(Array(labelIndexer, intPlanIndexer, assembler, xgb))

// Create a cross-validator.

import org.apache.spark.ml.tuning.CrossValidator

val cv = new CrossValidator()
         .setEstimator(pipeline)
         .setEvaluator(evaluator)
         .setEstimatorParamMaps(paramGrid)
         .setNumFolds(3)

// We can now fit the model using the training data. This will run
// cross-validation, choosing the best set of parameters.

val model = cv.fit(trainingData)
```

```
// You can now make some predictions on our test data.

val predictions = model.transform(testData)

predictions.printSchema

root
 |-- state: string (nullable = true)
 |-- account_length: double (nullable = true)
 |-- area_code: double (nullable = true)
 |-- phone_number: string (nullable = true)
 |-- international_plan: string (nullable = true)
 |-- voice_mail_plan: string (nullable = true)
 |-- number_vmail_messages: double (nullable = true)
 |-- total_day_minutes: double (nullable = true)
 |-- total_day_calls: double (nullable = true)
 |-- total_day_charge: double (nullable = true)
 |-- total_eve_minutes: double (nullable = true)
 |-- total_eve_calls: double (nullable = true)
 |-- total_eve_charge: double (nullable = true)
 |-- total_night_minutes: double (nullable = true)
 |-- total_night_calls: double (nullable = true)
 |-- total_night_charge: double (nullable = true)
 |-- total_intl_minutes: double (nullable = true)
 |-- total_intl_calls: double (nullable = true)
 |-- total_intl_charge: double (nullable = true)
 |-- number_customer_service_calls: double (nullable = true)
 |-- churned: string (nullable = true)
 |-- label: double (nullable = false)
 |-- int_plan: double (nullable = false)
 |-- features: vector (nullable = true)
 |-- rawPrediction: vector (nullable = true)
 |-- probability: vector (nullable = true)
 |-- prediction: double (nullable = false)
```

```
// Let's evaluate the model.

val auc = evaluator.evaluate(predictions)
auc: Double = 0.9328044307445879

// The AUC score produced by XGBoost4J-Spark is slightly better compared
// to our previous Random Forest example. XGBoost4J-Spark was also
// faster than Random Forest in training this dataset.

// Like Random Forest, XGBoost lets you extract the feature importance.

import ml.dmlc.xgboost4j.scala.spark.XGBoostClassificationModel
import org.apache.spark.ml.PipelineModel

val bestModel = model.bestModel

val model = bestModel
            .asInstanceOf[PipelineModel]
            .stages
            .last
            .asInstanceOf[XGBoostClassificationModel]

// Execute the getFeatureScore method to extract the feature importance.

model.nativeBooster.getFeatureScore()

res9: scala.collection.mutable.Map[String,Integer] = Map(f7 -> 4, f9 ->
7, f10 -> 2, f12 -> 4, f11 -> 8, f0 -> 5, f1 -> 19, f2 -> 17, f3 -> 10,
f4 -> 2, f5 -> 3)

// The method returns a map with the key mapping to the feature
// array index and the value corresponding to the feature importance score.
```

Table 3-2. Feature Importance Using XGBoost4J-Spark

Index	Feature	Feature Importance
0	number_customer_service_calls	2
1	total_day_minutes	15
2	total_eve_minutes	10
3	account_length	3
4	number_vmail_messages	2
5	total_day_calls	3
6	total_day_charge	Omitted
7	total_eve_calls	2
8	total_eve_charge	Omitted
9	total_night_calls	2
10	total_intl_calls	2
11	total_intl_charge	1
12	int_plan	5

Notice that the output is missing a couple of columns, specifically total_day_charge (f6) and total_eve_charge (f8). These are features that XGBoost has been deemed ineffective in improving the predictive accuracy of the model (see Table 3-2). Only features used in at least one split make it to the XGBoost feature importance output. There are several possible explanations. It could mean that the dropped features have very low or zero variance. It could also mean that these two features are highly correlated with other features.

There are some interesting things to note when comparing XGBoost's feature importance output to our previous Random Forest example. Note that while our previous Random Forest model considers number_customer_service_calls as one of the most important features, XGBoost ranks it as one of the least important. Similarly, the previous Random Forest model considers total_day_charge as the most important feature, but XGBoost completely omits it from the output due to its lack of importance (see Listing 3-6).

Listing 3-6. Extracting the Parameters of the XGBoost4J-Spark Model

```
val bestModel = model
                .bestModel
                .asInstanceOf[PipelineModel]
                .stages
                .last
                .asInstanceOf[XGBoostClassificationModel]

print(bestModel.extractParamMap)
{
        xgbc_9b95e70ab140-alpha: 0.0,
        xgbc_9b95e70ab140-baseScore: 0.5,
        xgbc_9b95e70ab140-checkpointInterval: -1,
        xgbc_9b95e70ab140-checkpointPath: ,
        xgbc_9b95e70ab140-colsampleBylevel: 1.0,
        xgbc_9b95e70ab140-colsampleBytree: 1.0,
        xgbc_9b95e70ab140-customEval: null,
        xgbc_9b95e70ab140-customObj: null,
        xgbc_9b95e70ab140-eta: 0.2,
        xgbc_9b95e70ab140-evalMetric: error,
        xgbc_9b95e70ab140-featuresCol: features,
        xgbc_9b95e70ab140-gamma: 0.0,
        xgbc_9b95e70ab140-growPolicy: depthwise,
        xgbc_9b95e70ab140-labelCol: label,
        xgbc_9b95e70ab140-lambda: 1.0,
        xgbc_9b95e70ab140-lambdaBias: 0.0,
        xgbc_9b95e70ab140-maxBin: 16,
        xgbc_9b95e70ab140-maxDeltaStep: 0.0,
        xgbc_9b95e70ab140-maxDepth: 8,
        xgbc_9b95e70ab140-minChildWeight: 1.0,
        xgbc_9b95e70ab140-missing: NaN,
        xgbc_9b95e70ab140-normalizeType: tree,
        xgbc_9b95e70ab140-nthread: 1,
        xgbc_9b95e70ab140-numEarlyStoppingRounds: 0,
```

```
        xgbc_9b95e70ab140-numRound: 1,
        xgbc_9b95e70ab140-numWorkers: 1,
        xgbc_9b95e70ab140-objective: reg:linear,
        xgbc_9b95e70ab140-predictionCol: prediction,
        xgbc_9b95e70ab140-probabilityCol: probability,
        xgbc_9b95e70ab140-rateDrop: 0.0,
        xgbc_9b95e70ab140-rawPredictionCol: rawPrediction,
        xgbc_9b95e70ab140-sampleType: uniform,
        xgbc_9b95e70ab140-scalePosWeight: 1.0,
        xgbc_9b95e70ab140-seed: 0,
        xgbc_9b95e70ab140-silent: 0,
        xgbc_9b95e70ab140-sketchEps: 0.03,
        xgbc_9b95e70ab140-skipDrop: 0.0,
        xgbc_9b95e70ab140-subsample: 1.0,
        xgbc_9b95e70ab140-timeoutRequestWorkers: 1800000,
        xgbc_9b95e70ab140-trackerConf: TrackerConf(0,python),
        xgbc_9b95e70ab140-trainTestRatio: 1.0,
        xgbc_9b95e70ab140-treeLimit: 0,
        xgbc_9b95e70ab140-treeMethod: auto,
        xgbc_9b95e70ab140-useExternalMemory: false
}
```

LightGBM: Fast Gradient Boosting from Microsoft

For years XGBoost has been everyone's favorite go-to algorithm for classification and regression. Lately, LightGBM has emerged as the new challenger to the throne. It is a relatively new tree-based gradient boosting variant similar to XGBoost. LightGBM was released on October 17, 2016, as part of Microsoft's Distributed Machine Learning Toolkit (DMTK) project. It was designed to be fast and distributed resulting in faster training speed and low memory usage. It supports GPU and parallel learning and the ability to handle large datasets. LightGBM has been shown in several benchmarks and experiments on public datasets to be even faster and with better accuracy than XGBoost.

Note LightGBM was ported to Spark as part of the Microsoft Machine Learning for Apache Spark (MMLSpark) ecosystem. Microsoft has been actively developing data science and deep learning tools with seamless integration with the Apache Spark ecosystem such as Microsoft Cognitive Toolkit, OpenCV, and LightGBM. MMLSpark requires Python 2.7 or 3.5+, Scala 2.11, and Spark 2.3+.

LightGBM has several advantages over XGBoost. It utilizes histograms to bucket continuous features into discrete bins. This provides LightGBM several performance advantages over XGBoost (which uses a pre-sort-based algorithm by default for tree learning) such as reduced memory usage, reduced cost of calculating the gain for each split, and reduced communication cost for parallel learning. LightGBM achieves additional performance boost by performing histogram subtraction on its sibling and parent to calculate a node's histogram. Benchmarks online show LightGBM is 11x to 15x faster than XGBoost (without binning) in some tasks.[xxix]

LightGBM generally outperforms XGBoost in terms of accuracy by growing trees leaf-wise (best-first). There are two main strategies for training decision trees, level-wise and leaf-wise (shown in Figure 3-7). Level-wise tree growth is the traditional way of growing decision trees for most tree-based ensembles (including XGBoost). LightGBM introduced the leaf-wise growth strategy. In contrast with level-wise growth, leaf-wise growth usually converges quicker[xxx] and achieves lower loss.[xxxi]

Note Leaf-wise growth tends to overfit with small datasets. It is advisable to set the max_depth parameter in LightGBM to limit tree depth. Note that even when max_depth is set, trees still grow leaf-wise.[xxxii] I discuss LightGBM parameter tuning later in the chapter.

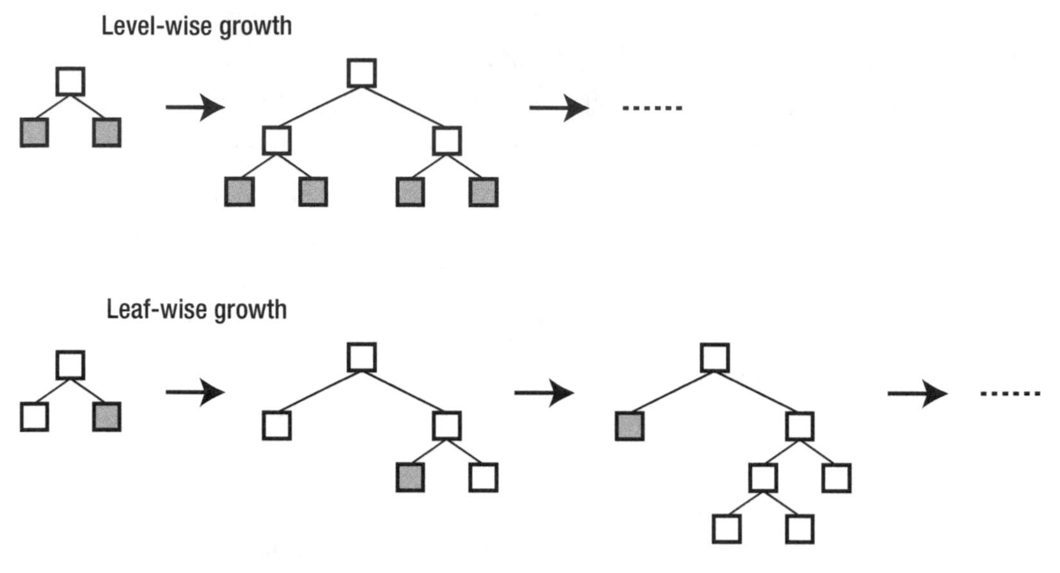

Figure 3-7. *Level-wise growth vs. leaf-wise growth*

Note XGBoost has since implemented many of the optimizations pioneered by LightGBM including leaf-wise tree growth strategy and using histograms to bucket continuous features into discrete bins. Latest benchmarks show XGBoost reaching performance competitive with LightGBM.[xxxiii]

Parameters

Tuning LightGBM is slightly more complicated compared to other algorithms such as Random Forest. LightGBM uses the leaf-wise (best-first) tree growth algorithm which can be susceptible to overfitting if parameters are not properly configured. Moreover, LightGBM has more than 100 parameters. Focusing on the most important parameters is enough to help you get started with LightGBM. You can learn the rest as you become more familiar with the algorithm.

- *max_depth*: Set this parameter to prevent trees from growing too deep. Shallow trees have less likelihood to overfit. Setting this parameter is particularly important if the dataset is small.

- *num_leaves*: Controls the complexity of the tree model. The value should be less than 2^(max_depth) to prevent overfitting. Setting num_leaves to a large value can increase accuracy at the risk of a higher chance of overfitting. Setting num_leaves to a small value can help prevent overfitting.
- *min_data_in_leaf:* Setting this parameter to a large value can prevent trees from growing too deep. This is another parameter you can set to help control overfitting. Setting the value too large can cause underfitting.
- *max_bin:* LightGBM groups the values of continuous features into discrete buckets using histograms. Set max_bin to specify the number of bins that the values will be grouped in. A small value can help control overfitting and improve training speed, while a larger value improves accuracy.
- *feature_fraction:* This parameter enables feature subsampling. This parameter specifies the fraction of features that will be randomly selected in each iteration. For instance, setting feature_fraction to 0.75 will randomly select 75% of the features in each iteration. Setting this parameter can increase training speed and help prevent overfitting.
- *bagging_fraction:* Specifies the fraction of data that will be selected in each iteration. For instance, setting bagging_fraction to 0.75 will randomly select 75% of the data in each iteration. Setting this parameter can increase training speed and help prevent overfitting.
- *num_iteration:* Sets the number of boosting iterations. The default is 100. For multiclass classification, LightGBM builds num_class * num_iterations trees. Setting this parameter influences training speed.
- *objective:* Like XGBoost, LightGBM supports multiple objectives. The default objective is set to regression. Set this parameter to specify the type of task that your model is trying to perform. For regression tasks, the options are regression_l2, regression_l1, poisson, quantile, mape, gamma, huber, fair, or tweedie. For classification tasks, the options are binary, multiclass, or multiclassova. It is important to set the objective correctly to avoid unpredictable results or poor accuracy.

As always, performing a parameter grid search to determine the optimal values for these parameters is highly recommended. For an exhaustive list of LightGBM parameters, consult the LightGBM online documentation.

Note LightGBM for Spark has not yet reached feature parity with LightGBM for Python as of this writing. While LightGBM for Spark includes the most important parameters, it is still missing a few. You can get a list of all the available parameters in LightGBM for Spark by visiting `https://bit.ly/2OqHl2M`. You can compare it with the complete list of LightGBM parameters at `https://bit.ly/3OYGyaO`.

Example

We will reuse the same telco churn dataset and most of the code from the previous Random Forest and XGBoost examples, shown in Listing 3-7.

Listing 3-7. Churn Prediction with LightGBM

```
spark-shell --packages Azure:mmlspark:0.15

// Load the CSV file into a DataFrame.

val dataDF = spark.read.format("csv")
             .option("header", "true")
             .option("inferSchema", "true")
             .load("churn_data.txt")

// Check the schema.

dataDF.printSchema

root
 |-- state: string (nullable = true)
 |-- account_length: double (nullable = true)
 |-- area_code: double (nullable = true)
 |-- phone_number: string (nullable = true)
 |-- international_plan: string (nullable = true)
```

```
 |-- voice_mail_plan: string (nullable = true)
 |-- number_vmail_messages: double (nullable = true)
 |-- total_day_minutes: double (nullable = true)
 |-- total_day_calls: double (nullable = true)
 |-- total_day_charge: double (nullable = true)
 |-- total_eve_minutes: double (nullable = true)
 |-- total_eve_calls: double (nullable = true)
 |-- total_eve_charge: double (nullable = true)
 |-- total_night_minutes: double (nullable = true)
 |-- total_night_calls: double (nullable = true)
 |-- total_night_charge: double (nullable = true)
 |-- total_intl_minutes: double (nullable = true)
 |-- total_intl_calls: double (nullable = true)
 |-- total_intl_charge: double (nullable = true)
 |-- number_customer_service_calls: double (nullable = true)
 |-- churned: string (nullable = true)

// Select a few columns.

dataDF.select("state","phone_number","international_plan","churned").show

+-----+------------+------------------+-------+
|state|phone_number|international_plan|churned|
+-----+------------+------------------+-------+
|   KS|    382-4657|                no|  False|
|   OH|    371-7191|                no|  False|
|   NJ|    358-1921|                no|  False|
|   OH|    375-9999|               yes|  False|
|   OK|    330-6626|               yes|  False|
|   AL|    391-8027|               yes|  False|
|   MA|    355-9993|                no|  False|
|   MO|    329-9001|               yes|  False|
|   LA|    335-4719|                no|  False|
|   WV|    330-8173|               yes|  False|
|   IN|    329-6603|                no|   True|
|   RI|    344-9403|                no|  False|
|   IA|    363-1107|                no|  False|
```

```
|   MT|    394-8006|                 no|  False|
|   IA|    366-9238|                 no|  False|
|   NY|    351-7269|                 no|   True|
|   ID|    350-8884|                 no|  False|
|   VT|    386-2923|                 no|  False|
|   VA|    356-2992|                 no|  False|
|   TX|    373-2782|                 no|  False|
+-----+------------+-----------------+-------+
only showing top 20 rows

import org.apache.spark.ml.feature.StringIndexer

val labelIndexer = new StringIndexer().setInputCol("churned").
setOutputCol("label")

val intPlanIndexer = new StringIndexer().setInputCol("international_plan").
setOutputCol("int_plan")

val features = Array("number_customer_service_calls","total_day_
minutes","total_eve_minutes","account_length","number_vmail_
messages","total_day_calls","total_day_charge","total_eve_calls","total_
eve_charge","total_night_calls","total_intl_calls","total_intl_
charge","int_plan")

import org.apache.spark.ml.feature.VectorAssembler

val assembler = new VectorAssembler()
                .setInputCols(features)
                .setOutputCol("features")

val seed = 1234

val Array(trainingData, testData) = dataDF.randomSplit(Array(0.9, 0.1), seed)

// Create a LightGBM classifier.

import com.microsoft.ml.spark.LightGBMClassifier

val lightgbm = new LightGBMClassifier()
               .setFeaturesCol("features")
               .setLabelCol("label")
```

```
               .setRawPredictionCol("rawPrediction")
               .setObjective("binary")

// Remember to set the correct objective using the setObjective method.
// Specifying the incorrect objective can affect accuracy or produce
// unpredictable results. In LightGBM the default objective is set to
// regression. In this example, we are performing binary classification, so
// we set the objective to binary.
```

Note Spark supports Barrier Execution Mode starting in version 2.4. LightGBM supports Barrier Execution Mode with the setUseBarrierExecutionMode method starting in version 0.18.

```
import org.apache.spark.ml.evaluation.BinaryClassificationEvaluator

val evaluator = new BinaryClassificationEvaluator()
                .setLabelCol("label")
                .setMetricName("areaUnderROC")

import org.apache.spark.ml.tuning.ParamGridBuilder

val paramGrid = new ParamGridBuilder()
                .addGrid(lightgbm.maxDepth, Array(2, 3, 4))
                .addGrid(lightgbm.numLeaves, Array(4, 6, 8))
                .addGrid(lightgbm.numIterations, Array(600))
                .build()

import org.apache.spark.ml.{ Pipeline, PipelineStage }

val pipeline = new Pipeline()
               .setStages(Array(labelIndexer, intPlanIndexer, assembler,
               lightgbm))

import org.apache.spark.ml.tuning.CrossValidator

val cv = new CrossValidator()
         .setEstimator(pipeline)
         .setEvaluator(evaluator)
```

```
        .setEstimatorParamMaps(paramGrid)
        .setNumFolds(3)

val model = cv.fit(trainingData)

// You can now make some predictions on our test data.

val predictions = model.transform(testData)

predictions.printSchema

root
 |-- state: string (nullable = true)
 |-- account_length: double (nullable = true)
 |-- area_code: double (nullable = true)
 |-- phone_number: string (nullable = true)
 |-- international_plan: string (nullable = true)
 |-- voice_mail_plan: string (nullable = true)
 |-- number_vmail_messages: double (nullable = true)
 |-- total_day_minutes: double (nullable = true)
 |-- total_day_calls: double (nullable = true)
 |-- total_day_charge: double (nullable = true)
 |-- total_eve_minutes: double (nullable = true)
 |-- total_eve_calls: double (nullable = true)
 |-- total_eve_charge: double (nullable = true)
 |-- total_night_minutes: double (nullable = true)
 |-- total_night_calls: double (nullable = true)
 |-- total_night_charge: double (nullable = true)
 |-- total_intl_minutes: double (nullable = true)
 |-- total_intl_calls: double (nullable = true)
 |-- total_intl_charge: double (nullable = true)
 |-- number_customer_service_calls: double (nullable = true)
 |-- churned: string (nullable = true)
 |-- label: double (nullable = false)
 |-- int_plan: double (nullable = false)
 |-- features: vector (nullable = true)
 |-- rawPrediction: vector (nullable = true)
 |-- probability: vector (nullable = true)
 |-- prediction: double (nullable = false)
```

```
// Evaluate the model. The AUC score is higher than Random Forest
// and XGBoost from our previous examples.

val auc = evaluator.evaluate(predictions)

auc: Double = 0.940366124260358

//LightGBM also lets you extract the feature importance.

import com.microsoft.ml.spark.LightGBMClassificationModel
import org.apache.spark.ml.PipelineModel

val bestModel = model.bestModel

val model = bestModel.asInstanceOf[PipelineModel]
            .stages
            .last
            .asInstanceOf[LightGBMClassificationModel]
```

There are two types of feature importance in LightGBM, "split" (total number of split) and "gain" (total information gain). Using "gain" is generally recommended and is roughly similar to Random Forest's method of calculating feature importance, but instead of using gini impurity, LightGBM uses cross entropy (log loss) in our binary classification example (see Tables 3-3 and 3-4). The loss to minimize depends on the specified objective.[xxxiv]

```
val gainFeatureImportances = model.getFeatureImportances("gain")

gainFeatureImportances: Array[Double] =
Array(2648.0893859118223, 5339.0795262902975, 2191.832309693098
,564.6461282968521, 1180.4672759771347, 656.8244850635529, 0.0,
533.6638155579567, 579.7435692846775, 651.5408382415771, 1179.492751300335,
2186.5995585918427, 1773.7864662855864)
```

Table 3-3. *Feature Importance with LightGBM Using Information Gain*

Index	Feature	Feature Importance
0	number_customer_service_calls	2648.0893859118223
1	total_day_minutes	5339.0795262902975
2	total_eve_minutes	2191.832309693098
3	account_length	564.6461282968521
4	number_vmail_messages	1180.4672759771347
5	total_day_calls	656.8244850635529
6	total_day_charge	0.0
7	total_eve_calls	533.6638155579567
8	total_eve_charge	579.7435692846775
9	total_night_calls	651.5408382415771
10	total_intl_calls	1179.492751300335
11	total_intl_charge	2186.5995585918427
12	int_plan	1773.7864662855864

Compare the output when using "split".

```
val gainFeatureImportances = model.getFeatureImportances("split")
gainFeatureImportances: Array[Double] = Array(159.0, 583.0, 421.0, 259.0,
133.0, 264.0, 0.0, 214.0, 92.0, 279.0, 279.0, 366.0, 58.0)
```

Table 3-4. *Feature Importance with LightGBM Using Number of Splits*

Index	Feature	Feature Importance
0	number_customer_service_calls	159.0
1	total_day_minutes	583.0
2	total_eve_minutes	421.0
3	account_length	259.0
4	number_vmail_messages	133.0
5	total_day_calls	264.0
6	total_day_charge	0.0
7	total_eve_calls	214.0
8	total_eve_charge	92.0
9	total_night_calls	279.0
10	total_intl_calls	279.0
11	total_intl_charge	366.0
12	int_plan	58.0

```
println(s"True Negative: ${predictions.select("*").where("prediction =
0 AND label = 0").count()}  True Positive: ${predictions.select("*").
where("prediction = 1 AND label = 1").count()}")

True Negative: 407  True Positive: 58

println(s"False Negative: ${predictions.select("*").where("prediction =
0 AND label = 1").count()} False Positive: ${predictions.select("*").
where("prediction = 1 AND label = 0").count()}")

False Negative: 20 False Positive: 9
```

Sentiment Analysis with Naïve Bayes

Naïve Bayes is a simple multiclass linear classification algorithm based on Bayes' theorem. Naïve Bayes got its name because it naively assumes that the features in a dataset are independent, ignoring any possible correlations between features. This is not the case in real-world scenarios, still Naïve Bayes tends to perform well especially on small datasets or datasets with high dimensionality. Like linear classifiers, it performs poorly on nonlinear classification problems. Naïve Bayes is a computationally efficient and highly scalable algorithm, only requiring a single pass to the dataset. It is a good baseline model for classification tasks using large datasets. It works by finding the probability that a point belongs to a class given a set of features. The Bayes' theorem equation can be stated as:

$$P(A|B) = \frac{P(B|A)P(A)}{P(B)}$$

P(A| B) is the *posterior probability* which can be interpreted as: "What is the probability of an event A happening given the event B?" with B representing the feature vectors. The numerator represents the *conditional probability* multiplied by the *prior probability*. The denominator represents the *evidence*. The equation can be more precisely written as:

$$P(y|x_1,\ldots,x_n) = \frac{P(x_1,\ldots,x_n|y)}{P(x_1,\ldots,x_n)}$$

Naïve Bayes is frequently used in text classification. Popular applications for text classification include spam detection and document classification. Another text classification use case is sentiment analysis. Companies regularly check comments from social media to determine if public opinion for a product or service is either positive or negative. Hedge funds use sentiment analysis to predict stock market movements.

Spark MLlib supports Bernoulli naïve Bayes and multinomial naïve Bayes. Bernoulli naïve Bayes only works with Boolean or binary features (e.g., the presence or absence of a word in the document), while multinomial naïve Bayes was designed for discrete features (e.g., word counts). The default model type for MLlib's naïve Bayes implementation is set to multinomial. You can set another parameter, lambda, for smoothing (the default is 1.0).

Example

Let's work on an example to demonstrate how we can use naïve Bayes for sentiment analysis. We will use a popular dataset from the University of California Irvine Machine Learning Repository. The dataset was created for the paper "From Group to Individual Labels using Deep Features," Kotzias et. al., KDD 2015. The dataset comes from three different companies: IMDB, Amazon, and Yelp. There are 500 positive and 500 negative reviews for each company. We will use the dataset from Amazon to determine the probability that the sentiment for a particular product is positive (1) or negative (0) based on Amazon product reviews.

We need to convert each sentence in the dataset into a feature vector. Spark MLlib provides a transformer for exactly this purpose. Term frequency–inverse document frequency (TF–IDF) is commonly used to generate feature vectors from texts. TF–IDF is used to determine the relevance of a word to a document in the corpus by computing the number of times the word occurs in the document (TF) and how frequently a word occurs across the whole corpus (IDF). In Spark MLlib, TF and IDF are implemented separately (HashingTF and IDF).

Before we can use TF–IDF to convert words into feature vectors, we need to use another transformer, tokenizer, to split the sentences into individual words. The steps should look like Figure 3-8, with the code shown in Listing 3-8.

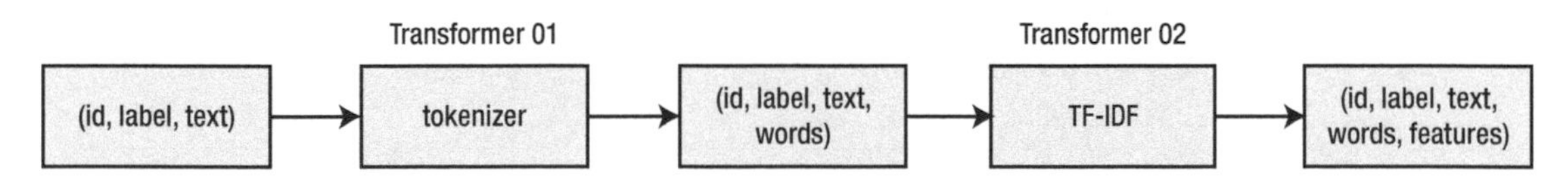

Figure 3-8. *Feature transformation for our sentiment analysis example*

Listing 3-8. Sentiment Analysis Using Naïve Bayes

```
// Start by creating a schema for our dataset.
import org.apache.spark.sql.types._

var reviewsSchema = StructType(Array (
    StructField("text",   StringType, true),
    StructField("label",  IntegerType, true)
    ))
```

```
// Create a DataFrame from the tab-delimited text file.
// Use the "csv" format regardless if its tab or comma delimited.
// The file does not have a header, so we'll set the header
// option to false. We'll set delimiter to tab and use the schema
// that we just built.

val reviewsDF = spark.read.format("csv")
                .option("header", "false")
                .option("delimiter","\t")
                .schema(reviewsSchema)
                .load("/files/amazon_cells_labelled.txt")

// Review the schema.

reviewsDF.printSchema

root
 |-- text: string (nullable = true)
 |-- label: integer (nullable = true)

// Check the data.

reviewsDF.show

+--------------------+-----+
|                text|label|
+--------------------+-----+
|So there is no wa...|    0|
|Good case, Excell...|    1|
|Great for the jaw...|    1|
|Tied to charger f...|    0|
|   The mic is great.|    1|
|I have to jiggle ...|    0|
|If you have sever...|    0|
|If you are Razr o...|    1|
|Needless to say, ...|    0|
|What a waste of m...|    0|
|And the sound qua...|    1|
|He was very impre...|    1|
```

```
|If the two were s...|    0|
|Very good quality...|    1|
|The design is ver...|    0|
|Highly recommend ...|    1|
|I advise EVERYONE...|    0|
|     So Far So Good!.|    1|
|        Works great!.|    1|
|It clicks into pl...|    0|
+--------------------+-----+
only showing top 20 rows

// Let's do some row counts.

reviewsDF.createOrReplaceTempView("reviews")

spark.sql("select label,count(*) from reviews group by label").show

+-----+--------+
|label|count(1)|
+-----+--------+
|    1|     500|
|    0|     500|
+-----+--------+

// Randomly divide the dataset into training and test datasets.

val seed = 1234

val Array(trainingData, testData) = reviewsDF.randomSplit(Array(0.8, 0.2), seed)

trainingData.count
res5: Long = 827

testData.count
res6: Long = 173

// Split the sentences into words.

import org.apache.spark.ml.feature.Tokenizer

val tokenizer = new Tokenizer().setInputCol("text")
                .setOutputCol("words")
```

```
// Check the tokenized data.

val tokenizedDF = tokenizer.transform(trainingData)

tokenizedDF.show

+--------------------+-----+--------------------+
|                text|label|               words|
+--------------------+-----+--------------------+
|         (It works!)|    1|      [(it, works!)]|
|)Setup couldn't h...|    1|[)setup, couldn't...|
|* Comes with a st...|    1|[*, comes, with, ...|
|.... Item arrived...|    1|[...., item, arri...|
|1. long lasting b...|    0|[1., long, lastin...|
|2 thumbs up to th...|    1|[2, thumbs, up, t...|
|:-)Oh, the charge...|    1|[:-)oh,, the, cha...|
|   A Disappointment.|    0|[a, disappointment.]|
|A PIECE OF JUNK T...|    0|[a, piece, of, ju...|
|A good quality ba...|    1|[a, good, quality...|
|A must study for ...|    0|[a, must, study, ...|
|A pretty good pro...|    1|[a, pretty, good,...|
|A usable keyboard...|    1|[a, usable, keybo...|
|A week later afte...|    0|[a, week, later, ...|
|AFTER ARGUING WIT...|    0|[after, arguing, ...|
|AFter the first c...|    0|[after, the, firs...|
|       AMAZON SUCKS.|    0|     [amazon, sucks.]|
|     Absolutel junk.|    0|  [absolutel, junk.]|
|   Absolutely great.|    1|[absolutely, great.]|
|Adapter does not ...|    0|[adapter, does, n...|
+--------------------+-----+--------------------+
only showing top 20 rows

// Next, we'll use HashingTF to convert the tokenized words
// into fixed-length feature vector.

import org.apache.spark.ml.feature.HashingTF

val htf = new HashingTF().setNumFeatures(1000)
          .setInputCol("words")
```

```
.setOutputCol("features")

// Check the vectorized features.

val hashedDF = htf.transform(tokenizedDF)

hashedDF.show

+--------------------+-----+--------------------+--------------------+
|                text|label|               words|            features|
+--------------------+-----+--------------------+--------------------+
|         (It works!)|    1|       [(it, works!)]|(1000,[369,504],[...|
|)Setup couldn't h...|    1|[)setup, couldn't...|(1000,[299,520,53...|
|* Comes with a st...|    1|[*, comes, with, ...|(1000,[34,51,67,1...|
|.... Item arrived...|    1|[...., item, arri...|(1000,[98,133,245...|
|1. long lasting b...|    0|[1., long, lastin...|(1000,[138,258,29...|
|2 thumbs up to th...|    1|[2, thumbs, up, t...|(1000,[92,128,373...|
|:-)Oh, the charge...|    1|[:-)oh,, the, cha...|(1000,[388,497,52...|
|   A Disappointment.|    0|[a, disappointment.]|(1000,[170,386],[...|
|A PIECE OF JUNK T...|    0|[a, piece, of, ju...|(1000,[34,36,47,7...|
|A good quality ba...|    1|[a, good, quality...|(1000,[77,82,168,...|
|A must study for ...|    0|[a, must, study, ...|(1000,[23,36,104,...|
|A pretty good pro...|    1|[a, pretty, good,...|(1000,[168,170,27...|
|A usable keyboard...|    1|[a, usable, keybo...|(1000,[2,116,170,...|
|A week later afte...|    0|[a, week, later, ...|(1000,[77,122,156...|
|AFTER ARGUING WIT...|    0|[after, arguing, ...|(1000,[77,166,202...|
|AFter the first c...|    0|[after, the, firs...|(1000,[63,77,183,...|
|       AMAZON SUCKS.|    0|    [amazon, sucks.]|(1000,[828,966],[...|
|     Absolutel junk.|    0|  [absolutel, junk.]|(1000,[607,888],[...|
|   Absolutely great.|    1|[absolutely, great.]|(1000,[589,903],[...|
|Adapter does not ...|    0|[adapter, does, n...|(1000,[0,18,51,28...|
+--------------------+-----+--------------------+--------------------+
only showing top 20 rows

// We will use the naïve Bayes classifier provided by MLlib.

import org.apache.spark.ml.classification.NaiveBayes

val nb = new NaiveBayes()
```

```
// We now have all the parts that we need to assemble
// a machine learning pipeline.

import org.apache.spark.ml.Pipeline

val pipeline = new Pipeline().setStages(Array(tokenizer, htf, nb))

// Train our model using the training dataset.

val model = pipeline.fit(trainingData)

// Predict using the test dataset.

val predictions = model.transform(testData)

// Display the predictions for each review.

predictions.select("text","prediction").show

+--------------------+----------+
|                text|prediction|
+--------------------+----------+
|!I definitely reco...|       1.0|
|#1 It Works - #2 ...|       1.0|
| $50 Down the drain.|       0.0|
|A lot of websites...|       1.0|
|After charging ov...|       0.0|
|After my phone go...|       0.0|
|All in all I thin...|       1.0|
|All it took was o...|       0.0|
|Also, if your pho...|       0.0|
|And I just love t...|       1.0|
|And none of the t...|       1.0|
|         Bad Choice.|       0.0|
|Best headset ever...|       1.0|
|Big Disappointmen...|       0.0|
|Bluetooth range i...|       0.0|
|But despite these...|       0.0|
|Buyer--Be Very Ca...|       1.0|
|Can't store anyth...|       0.0|
```

```
|Chinese Forgeries...|       0.0|
|Do NOT buy if you...|       0.0|
+--------------------+----------+

only showing top 20 rows

// Evaluate our model using a binary classifier evaluator.

import org.apache.spark.ml.evaluation.BinaryClassificationEvaluator

val evaluator = new BinaryClassificationEvaluator()

import org.apache.spark.ml.param.ParamMap

val paramMap = ParamMap(evaluator.metricName -> "areaUnderROC")

val auc = evaluator.evaluate(predictions, paramMap)

auc: Double = 0.5407085561497325

// Test on a positive example.

val predictions = model
.transform(sc.parallelize(Seq("This product is good")).toDF("text"))

predictions.select("text","prediction").show

+--------------------+----------+
|                text|prediction|
+--------------------+----------+
|This product is good|       1.0|
+--------------------+----------+

// Test on a negative example.

val predictions = model
.transform(sc.parallelize(Seq("This product is bad")).toDF("text"))

predictions.select("text","prediction").show

+-------------------+----------+
|               text|prediction|
+-------------------+----------+
|This product is bad|       0.0|
+-------------------+----------+
```

There are several things that can be done to improve our model. Performing additional text preprocessing such as n-grams, lemmatization, and removal of stop words are common in most natural language processing (NLP) tasks. I cover Stanford CoreNLP and Spark NLP in Chapter 4.

Regression

Regression is a supervised machine learning task for predicting continuous numeric values. Popular use cases include sales and demand forecasting, predicting stock, home or commodity prices, and weather forecasting, to mention a few. I discuss regression in more detail in Chapter 1.

Simple Linear Regression

Linear regression is used for examining linear relationships between one or more independent variable(s) and a dependent variable. The analysis of the relationship between a single independent variable and a single continuous dependent variable is known as simple linear regression.

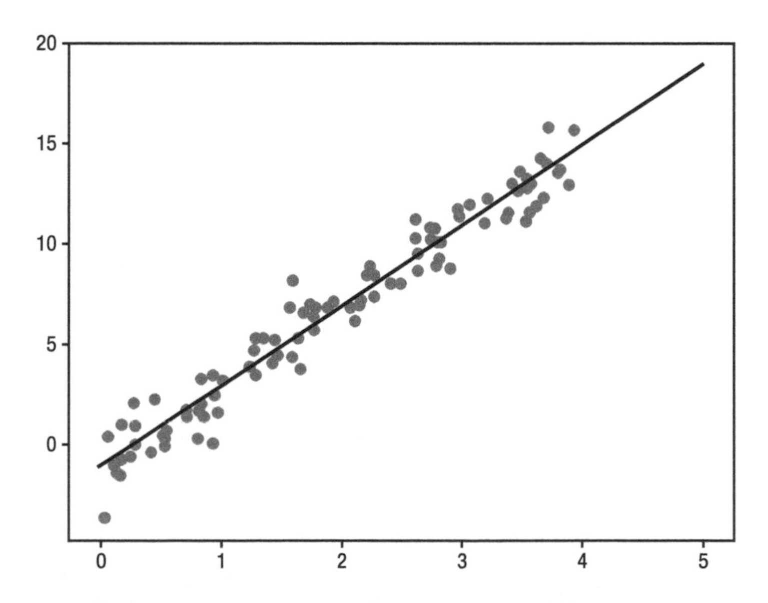

Figure 3-9. *A simple linear regression plot*

As you can see in Figure 3-9, the plot shows linear aggression trying to draw a straight line that will best reduce the residual sum of squares between the observed responses and the predicted values.[xxxv]

Example

For our example, we will use simple linear regression to show how home prices (the dependent variable) change based on the area's average household income (the independent variable). Listing 3-9 details the code.

Listing 3-9. Linear Regression Example

```
import org.apache.spark.ml.regression.LinearRegression
import spark.implicits._

val dataDF = Seq(
 (50000, 302200),
 (75200, 550000),
 (90000, 680000),
 (32800, 225000),
 (41000, 275000),
 (54000, 300500),
 (72000, 525000),
 (105000, 700000),
 (88500, 673100),
 (92000, 695000),
 (53000, 320900),
 (85200, 652800),
 (157000, 890000),
 (128000, 735000),
 (71500, 523000),
 (114000, 720300),
 (33400, 265900),
 (143000, 846000),
 (68700, 492000),
 (46100, 285000)
).toDF("avg_area_income","price")

dataDF.show
```

```
+---------------+------+
|avg_area_income| price|
+---------------+------+
|          50000|302200|
|          75200|550000|
|          90000|680000|
|          32800|225000|
|          41000|275000|
|          54000|300500|
|          72000|525000|
|         105000|700000|
|          88500|673100|
|          92000|695000|
|          53000|320900|
|          85200|652800|
|         157000|890000|
|         128000|735000|
|          71500|523000|
|         114000|720300|
|          33400|265900|
|         143000|846000|
|          68700|492000|
|          46100|285000|
+---------------+------+

import org.apache.spark.ml.feature.VectorAssembler

val assembler = new VectorAssembler()
                .setInputCols(Array("avg_area_income"))
                .setOutputCol("feature")

val dataDF2 = assembler.transform(dataDF)

dataDF2.show
```

```
+---------------+------+----------+
|avg_area_income| price|   feature|
+---------------+------+----------+
|          50000|302200| [50000.0]|
|          75200|550000| [75200.0]|
|          90000|680000| [90000.0]|
|          32800|225000| [32800.0]|
|          41000|275000| [41000.0]|
|          54000|300500| [54000.0]|
|          72000|525000| [72000.0]|
|         105000|700000|[105000.0]|
|          88500|673100| [88500.0]|
|          92000|695000| [92000.0]|
|          53000|320900| [53000.0]|
|          85200|652800| [85200.0]|
|         157000|890000|[157000.0]|
|         128000|735000|[128000.0]|
|          71500|523000| [71500.0]|
|         114000|720300|[114000.0]|
|          33400|265900| [33400.0]|
|         143000|846000|[143000.0]|
|          68700|492000| [68700.0]|
|          46100|285000| [46100.0]|
+---------------+------+----------+

val lr = new LinearRegression()
         .setMaxIter(10)
         .setFeaturesCol("feature")
         .setLabelCol("price")

val model = lr.fit(dataDF2)

import org.apache.spark.ml.linalg.Vectors

val testData = spark
         .createDataFrame(Seq(Vectors.dense(75000))
         .map(Tuple1.apply))
         .toDF("feature")
```

```
val predictions = model.transform(testData)

predictions.show
+---------+------------------+
|  feature|        prediction|
+---------+------------------+
|[75000.0]|504090.35842779215|
+---------+------------------+
```

Multiple Regression with XGBoost4J-Spark

Multiple regression is used in more realistic scenarios where there are two or more independent variables and a single continuous dependent variable. It is common to have both linear and nonlinear features in real-world use cases. Tree-based ensemble algorithms like XGBoost have the ability to handle both linear and nonlinear features which makes it ideal for most production environments. In most situations using tree-based ensembles such as XGBoost for multiple regression should lead to significantly better prediction accuracy.[xxxvi]

We used XGBoost to solve a classification problem earlier in the chapter. Because XGBoost supports both classification and regression, using XGBoost for regression is very similar to classification.

Example

We will use a slightly more complex dataset for our multiple regression example, shown in Listing 3-10. The dataset can be downloaded from Kaggle.[xxxvii] Our goal is to predict home prices based on the attributes provided in the dataset. The dataset contains seven columns: Avg. Area Income, Avg. Area House Age, Avg. Area Number of Room, Avg. Area Number of Bedrooms, Area Population, Price, and Address. We will not use the Address field for simplicity's sake (useful information can be derived from a home address such the location of nearby schools). Price is our dependent variable.

Listing 3-10. Multiple Regression Using XGBoost4J-Spark

```
spark-shell --packages ml.dmlc:xgboost4j-spark:0.81

import org.apache.spark.sql.types._

// Define a schema for our dataset.
```

```
var pricesSchema = StructType(Array (
    StructField("avg_area_income",   DoubleType, true),
    StructField("avg_area_house_age",   DoubleType, true),
    StructField("avg_area_num_rooms",   DoubleType, true),
    StructField("avg_area_num_bedrooms",   DoubleType, true),
    StructField("area_population",   DoubleType, true),
    StructField("price",   DoubleType, true)
    ))

val dataDF = spark.read.format("csv")
             .option("header","true")
             .schema(pricesSchema)
             .load("USA_Housing.csv").na.drop()

// Inspect the dataset.

dataDF.printSchema
root
 |-- avg_area_income: double (nullable = true)
 |-- avg_area_house_age: double (nullable = true)
 |-- avg_area_num_rooms: double (nullable = true)
 |-- avg_area_num_bedrooms: double (nullable = true)
 |-- area_population: double (nullable = true)
 |-- price: double (nullable = true)

dataDF.select("avg_area_income","avg_area_house_age","avg_area_num_rooms").
show

+-----------------+------------------+------------------+
|   avg_area_income|avg_area_house_age|avg_area_num_rooms|
+-----------------+------------------+------------------+
| 79545.45857431678| 5.682861321615587| 7.009188142792237|
| 79248.64245482568|6.0028998082752425| 6.730821019094919|
|61287.067178656784| 5.865889840310001| 8.512727430375099|
| 63345.24004622798|7.1882360945186425| 5.586728664827653|
|59982.197225708034| 5.040554523106283| 7.839387785120487|
|  80175.7541594853|4.9884077575337145| 6.104512439428879|
| 64698.46342788773| 6.025335906887153| 8.147759585023431|
```

```
| 78394.33927753085|6.9897797477182815| 6.620477995185026|
| 59927.66081334963|  5.36212556960358|6.3931209805509015|
| 81885.92718409566| 4.423671789897876| 8.167688003472351|
| 80527.47208292288|  8.09351268063935| 5.042746799645982|
| 50593.69549704281| 4.496512793097035| 7.467627404008019|
|39033.809236982364| 7.671755372854428| 7.250029317273495|
|  73163.6634410467| 6.919534825456555|5.9931879009455695|
|  69391.3801843616| 5.344776176735725| 8.406417714534253|
| 73091.86674582321| 5.443156466535474| 8.517512711137975|
| 79706.96305765743| 5.067889591058972| 8.219771123286257|
| 61929.07701808926| 4.788550241805888|5.0970095543775615|
| 63508.19429942997| 5.947165139552473| 7.187773835329727|
| 62085.27640340488| 5.739410843630574|  7.09180810424997|
+------------------+------------------+------------------+
only showing top 20 rows

dataDF.select("avg_area_num_bedrooms","area_population","price").show

+---------------------+------------------+------------------+
|avg_area_num_bedrooms|   area_population|             price|
+---------------------+------------------+------------------+
|                 4.09|23086.800502686456|1059033.5578701235|
|                 3.09| 40173.07217364482|  1505890.91484695|
|                 5.13| 36882.15939970458|1058987.9878760849|
|                 3.26| 34310.24283090706|1260616.8066294468|
|                 4.23|26354.109472103148| 630943.4893385402|
|                 4.04|26748.428424689715|1068138.0743935304|
|                 3.41| 60828.24908540716|1502055.8173744078|
|                 2.42|36516.358972493836|1573936.5644777215|
|                  2.3| 29387.39600281585| 798869.5328331633|
|                  6.1| 40149.96574921337|1545154.8126419624|
|                  4.1| 47224.35984022191|  1707045.722158058|
|                 4.49|34343.991885578806| 663732.3968963273|
|                  3.1| 39220.36146737246|1042814.0978200927|
|                 2.27|32326.123139488096|1291331.5184858206|
|                 4.37|35521.294033173246|1402818.2101658515|
|                 4.01|23929.524053267953|1306674.6599511993|
```

```
|                3.12| 39717.81357630952|1556786.6001947748|
|                 4.3| 24595.90149782299| 528485.2467305964|
|                5.12|35719.653052030866|1019425.9367578316|
|                5.49|44922.106702293066|1030591.4292116085|
+--------------------+------------------+------------------+
only showing top 20 rows

val features = Array("avg_area_income","avg_area_house_age",
"avg_area_num_rooms","avg_area_num_bedrooms","area_population")

// Combine our features into a single feature vector.

import org.apache.spark.ml.feature.VectorAssembler

val assembler = new VectorAssembler()
                .setInputCols(features)
                .setOutputCol("features")

val dataDF2 = assembler.transform(dataDF)

dataDF2.select("price","features").show(20,50)

+------------------+--------------------------------------------------+
|             price|                                          features|
+------------------+--------------------------------------------------+
|1059033.5578701235|[79545.45857431678,5.682861321615587,7.00918814...|
|  1505890.91484695|[79248.64245482568,6.0028998082752425,6.7308210...|
|1058987.9878760849|[61287.067178656784,5.865889840310001,8.5127274...|
|1260616.8066294468|[63345.24004622798,7.1882360945186425,5.5867286...|
| 630943.4893385402|[59982.197225708034,5.040554523106283,7.8393877...|
|1068138.0743935304|[80175.7541594853,4.9884077575337145,6.10451243...|
|1502055.8173744078|[64698.46342788773,6.025335906887153,8.14775958...|
|1573936.5644777215|[78394.33927753085,6.9897797477182815,6.6204779...|
| 798869.5328331633|[59927.66081334963,5.3621255696035 8,6.393120980...|
|1545154.8126419624|[81885.92718409566,4.423671789897876,8.16768800...|
| 1707045.722158058|[80527.47208292288,8.09351268063935,5.042746799...|
| 663732.3968963273|[50593.69549704281,4.496512793097035,7.46762740...|
|1042814.0978200927|[39033.809236982364,7.671755372854428,7.2500293...|
|1291331.5184858206|[73163.6634410467,6.919534825456555,5.993187900...|
```

```
|1402818.2101658515|[69391.3801843616,5.344776176735725,8.406417714...|
|1306674.6599511993|[73091.86674582321,5.443156466535474,8.51751271...|
|1556786.6001947748|[79706.96305765743,5.067889591058972,8.21977112...|
| 528485.2467305964|[61929.07701808926,4.788550241805888,5.09700955...|
|1019425.9367578316|[63508.19429942997,5.947165139552473,7.18777383...|
|1030591.4292116085|[62085.27640340488,5.739410843630574,7.09180810...|
+------------------+--------------------------------------------------+
only showing top 20 rows

// Divide our dataset into training and test data.

val seed = 1234

val Array(trainingData, testData) = dataDF2.randomSplit(Array(0.8, 0.2), seed)

// Use XGBoost for regression.

import ml.dmlc.xgboost4j.scala.spark.{XGBoostRegressionModel,XGBoostRegressor}

val xgb = new XGBoostRegressor()
          .setFeaturesCol("features")
          .setLabelCol("price")

// Create a parameter grid.

import org.apache.spark.ml.tuning.ParamGridBuilder

val paramGrid = new ParamGridBuilder()
                .addGrid(xgb.maxDepth, Array(6, 9))
                .addGrid(xgb.eta, Array(0.3, 0.7)).build()

paramGrid: Array[org.apache.spark.ml.param.ParamMap] =
Array({
      xgbr_bacf108db722-eta: 0.3,
      xgbr_bacf108db722-maxDepth: 6
}, {
      xgbr_bacf108db722-eta: 0.3,
      xgbr_bacf108db722-maxDepth: 9
}, {
      xgbr_bacf108db722-eta: 0.7,
      xgbr_bacf108db722-maxDepth: 6
```

```
}, {
      xgbr_bacf108db722-eta: 0.7,
      xgbr_bacf108db722-maxDepth: 9
})

// Create our evaluator.

import org.apache.spark.ml.evaluation.RegressionEvaluator

val evaluator = new RegressionEvaluator()
               .setLabelCol("price")
               .setPredictionCol("prediction")
               .setMetricName("rmse")

// Create our cross-validator.

import org.apache.spark.ml.tuning.CrossValidator

val cv = new CrossValidator()
         .setEstimator(xgb)
         .setEvaluator(evaluator)
         .setEstimatorParamMaps(paramGrid)
         .setNumFolds(3)

val model = cv.fit(trainingData)

val predictions = model.transform(testData)

predictions.select("features","price","prediction").show

+--------------------+------------------+------------+
|            features|             price|  prediction|
+--------------------+------------------+------------+
|[17796.6311895433...|302355.83597895555| 591896.9375|
|[35454.7146594754...| 1077805.577726322|   440094.75|
|[35608.9862370775...| 449331.5835333807|   672114.75|
|[38868.2503114142...| 759044.6879907805|   672114.75|
|[40752.7142433209...| 560598.5384309639| 591896.9375|
|[41007.4586732745...| 494742.5435776913|421605.28125|
|[41533.0129597444...| 682200.3005599922|505685.96875|
|[42258.7745410484...| 852703.2636757497| 591896.9375|
```

```
|[42940.1389392421...| 680418.7240122693| 591896.9375|
|[43192.1144092488...|1054606.9845532854|505685.96875|
|[43241.9824225005...| 629657.6132544072|505685.96875|
|[44328.2562966742...| 601007.3511604669|141361.53125|
|[45347.1506816944...| 541953.9056802422|441908.40625|
|[45546.6434075757...|   923830.33486809| 591896.9375|
|[45610.9384142094...|  961354.287727855|   849175.75|
|[45685.2499205068...| 867714.3838490517|441908.40625|
|[45990.1237417814...|1043968.3994445396|   849175.75|
|[46062.7542664558...| 675919.6815570832|505685.96875|
|[46367.2058588838...|268050.81474351394|  379889.625|
|[47467.4239151893...| 762144.9261238109| 591896.9375|
+--------------------+------------------+------------+
only showing top 20 rows
```

Let's evaluate the model using the root mean squared error (RMSE). Residuals are a measure of the distance of the data points from the regression line. The RMSE is the standard deviation of the residuals and is used to measure the prediction error.[xxxviii]

```
val rmse = evaluator.evaluate(predictions)

rmse: Double = 438499.82356536255

// Extract the parameters.

model.bestModel.extractParamMap

res11: org.apache.spark.ml.param.ParamMap =
{
      xgbr_8da6032c61a9-alpha: 0.0,
      xgbr_8da6032c61a9-baseScore: 0.5,
      xgbr_8da6032c61a9-checkpointInterval: -1,
      xgbr_8da6032c61a9-checkpointPath: ,
      xgbr_8da6032c61a9-colsampleBylevel: 1.0,
      xgbr_8da6032c61a9-colsampleBytree: 1.0,
      xgbr_8da6032c61a9-customEval: null,
      xgbr_8da6032c61a9-customObj: null,
      xgbr_8da6032c61a9-eta: 0.7,
      xgbr_8da6032c61a9-evalMetric: rmse,
```

```
        xgbr_8da6032c61a9-featuresCol: features,
        xgbr_8da6032c61a9-gamma: 0.0,
        xgbr_8da6032c61a9-growPolicy: depthwise,
        xgbr_8da6032c61a9-labelCol: price,
        xgbr_8da6032c61a9-lambda: 1.0,
        xgbr_8da6032c61a9-lambdaBias: 0.0,
        xgbr_8da6032c61a9-maxBin: 16,
        xgbr_8da6032c61a9-maxDeltaStep: 0.0,
        xgbr_8da6032c61a9-maxDepth: 9,
        xgbr_8da6032c61a9-minChildWeight: 1.0,
        xgbr_8da6032c61a9-missing: NaN,
        xgbr_8da6032c61a9-normalizeType: tree,
        xgbr_8da6032c61a9-nthread: 1,
        xgbr_8da6032c61a9-numEarlyStoppingRounds: 0,
        xgbr_8da6032c61a9-numRound: 1,
        xgbr_8da6032c61a9-numWorkers: 1,
        xgbr_8da6032c61a9-objective: reg:linear,
        xgbr_8da6032c61a9-predictionCol: prediction,
        xgbr_8da6032c61a9-rateDrop: 0.0,
        xgbr_8da6032c61a9-sampleType: uniform,
        xgbr_8da6032c61a9-scalePosWeight: 1.0,
        xgbr_8da6032c61a9-seed: 0,
        xgbr_8da6032c61a9-silent: 0,
        xgbr_8da6032c61a9-sketchEps: 0.03,
        xgbr_8da6032c61a9-skipDrop: 0.0,
        xgbr_8da6032c61a9-subsample: 1.0,
        xgbr_8da6032c61a9-timeoutRequestWorkers: 1800000,
        xgbr_8da6032c61a9-trackerConf: TrackerConf(0,python),
        xgbr_8da6032c61a9-trainTestRatio: 1.0,
        xgbr_8da6032c61a9-treeLimit: 0,
        xgbr_8da6032c61a9-treeMethod: auto,
        xgbr_8da6032c61a9-useExternalMemory: false
}
```

Multiple Regression with LightGBM

In Listing 3-11, we will use LightGBM. LightGBM comes with the LightGBMRegressor class specifically for regression tasks. We will reuse the housing dataset and most of the code from our previous XGBoost example.

Listing 3-11. Multiple Regression Using LightGBM

```
spark-shell --packages Azure:mmlspark:0.15

var pricesSchema = StructType(Array (
    StructField("avg_area_income",   DoubleType, true),
    StructField("avg_area_house_age",   DoubleType, true),
    StructField("avg_area_num_rooms",   DoubleType, true),
    StructField("avg_area_num_bedrooms",   DoubleType, true),
    StructField("area_population",   DoubleType, true),
    StructField("price",   DoubleType, true)
    ))

val dataDF = spark.read.format("csv")
             .option("header","true")
             .schema(pricesSchema)
             .load("USA_Housing.csv")
             .na.drop()

dataDF.printSchema

root
 |-- avg_area_income: double (nullable = true)
 |-- avg_area_house_age: double (nullable = true)
 |-- avg_area_num_rooms: double (nullable = true)
 |-- avg_area_num_bedrooms: double (nullable = true)
 |-- area_population: double (nullable = true)
 |-- price: double (nullable = true)
```

```
dataDF.select("avg_area_income","avg_area_house_age",
"avg_area_num_rooms")
.show
+-----------------+------------------+------------------+
|   avg_area_income|avg_area_house_age|avg_area_num_rooms|
+-----------------+------------------+------------------+
| 79545.45857431678| 5.682861321615587| 7.009188142792237|
| 79248.64245482568|6.0028998082752425| 6.730821019094919|
|61287.067178656784| 5.865889840310001| 8.512727430375099|
| 63345.24004622798|7.1882360945186425| 5.586728664827653|
|59982.197225708034| 5.040554523106283| 7.839387785120487|
|  80175.7541594853|4.9884077575337145| 6.104512439428879|
| 64698.46342788773| 6.025335906887153| 8.147759585023431|
| 78394.33927753085|6.9897797477182815| 6.620477995185026|
| 59927.66081334963|  5.36212556960358|6.3931209805509015|
| 81885.92718409566| 4.423671789897876| 8.167688003472351|
| 80527.47208292288|  8.09351268063935| 5.042746799645982|
| 50593.69549704281| 4.496512793097035| 7.467627404008019|
|39033.809236982364| 7.671755372854428| 7.250029317273495|
|  73163.6634410467| 6.919534825456555|5.9931879009455695|
|  69391.3801843616| 5.344776176735725| 8.406417714534253|
| 73091.86674582321| 5.443156466535474| 8.517512711137975|
| 79706.96305765743| 5.067889591058972| 8.219771123286257|
| 61929.07701808926| 4.788550241805888|5.0970095543775615|
| 63508.19429942997| 5.947165139552473| 7.187773835329727|
| 62085.27640340488| 5.739410843630574|  7.09180810424997|
+-----------------+------------------+------------------+

dataDF.select("avg_area_num_bedrooms","area_population","price").show

+---------------------+------------------+------------------+
|avg_area_num_bedrooms|   area_population|             price|
+---------------------+------------------+------------------+
|                 4.09|23086.800502686456|1059033.5578701235|
|                 3.09| 40173.07217364482|  1505890.91484695|
|                 5.13| 36882.15939970458|1058987.9878760849|
```

```
|                  3.26| 34310.24283090706|1260616.8066294468|
|                  4.23|26354.109472103148| 630943.4893385402|
|                  4.04|26748.428424689715|1068138.0743935304|
|                  3.41| 60828.24908540716|1502055.8173744078|
|                  2.42|36516.358972493836|1573936.5644777215|
|                   2.3| 29387.39600281585| 798869.5328331633|
|                   6.1| 40149.96574921337|1545154.8126419624|
|                   4.1| 47224.35984022191| 1707045.722158058|
|                  4.49|34343.991885578806| 663732.3968963273|
|                   3.1| 39220.36146737246|1042814.0978200927|
|                  2.27|32326.123139488096|1291331.5184858206|
|                  4.37|35521.294033173246|1402818.2101658515|
|                  4.01|23929.524053267953|1306674.6599511993|
|                  3.12| 39717.81357630952|1556786.6001947748|
|                   4.3| 24595.90149782299| 528485.2467305964|
|                  5.12|35719.653052030866|1019425.9367578316|
|                  5.49|44922.106702293066|1030591.4292116085|
+----------------------+------------------+------------------+
only showing top 20 rows

val features = Array("avg_area_income","avg_area_house_age",
"avg_area_num_rooms","avg_area_num_bedrooms","area_population")

import org.apache.spark.ml.feature.VectorAssembler

val assembler = new VectorAssembler()
                .setInputCols(features)
                .setOutputCol("features")

val dataDF2 = assembler.transform(dataDF)

dataDF2.select("price","features").show(20,50)

+------------------+--------------------------------------------------+
|             price|                                          features|
+------------------+--------------------------------------------------+
|1059033.5578701235|[79545.45857431678,5.682861321615587,7.00918814...|
|  1505890.91484695|[79248.64245482568,6.0028998082752425,6.7308210...|
|1058987.9878760849|[61287.067178656784,5.865889840310001,8.5127274...|
```

```
|1260616.8066294468|[63345.24004622798,7.1882360945186425,5.5867286...|
| 630943.4893385402|[59982.197225708034,5.040554523106283,7.8393877...|
|1068138.0743935304|[80175.7541594853,4.9884077575337145,6.10451243...|
|1502055.8173744078|[64698.46342788773,6.025335906887153,8.14775958...|
|1573936.5644777215|[78394.33927753085,6.9897797477182815,6.6204779...|
| 798869.5328331633|[59927.66081334963,5.3621255696035 8,6.393120980...|
|1545154.8126419624|[81885.92718409566,4.423671789897876,8.16768800...|
| 1707045.722158058|[80527.47208292288,8.09351268063935,5.042746799...|
| 663732.3968963273|[50593.69549704281,4.496512793097035,7.46762740...|
|1042814.0978200927|[39033.809236982364,7.671755372854428,7.2500293...|
|1291331.5184858206|[73163.6634410467,6.919534825456555,5.993187900...|
|1402818.2101658515|[69391.3801843616,5.344776176735725,8.406417714...|
|1306674.6599511993|[73091.86674582321,5.443156466535474,8.51751271...|
|1556786.6001947748|[79706.96305765743,5.067889591058972,8.21977112...|
| 528485.2467305964|[61929.07701808926,4.788550241805888,5.09700955...|
|1019425.9367578316|[63508.19429942997,5.947165139552473,7.18777383...|
|1030591.4292116085|[62085.27640340488,5.739410843630574,7.09180810...|
+------------------+--------------------------------------------------+
only showing top 20 rows

val seed = 1234

val Array(trainingData, testData) = dataDF2.randomSplit(Array(0.8, 0.2), seed)

import com.microsoft.ml.spark.{LightGBMRegressionModel,LightGBMRegressor}

val lightgbm = new LightGBMRegressor()
               .setFeaturesCol("features")
               .setLabelCol("price")
               .setObjective("regression")

import org.apache.spark.ml.tuning.ParamGridBuilder

val paramGrid = new ParamGridBuilder()
                .addGrid(lightgbm.numLeaves, Array(6, 9))
                .addGrid(lightgbm.numIterations, Array(10, 15))
                .addGrid(lightgbm.maxDepth, Array(2, 3, 4))
                .build()
```

```
paramGrid: Array[org.apache.spark.ml.param.ParamMap] =
Array({
        LightGBMRegressor_f969f7c475b5-maxDepth: 2,
        LightGBMRegressor_f969f7c475b5-numIterations: 10,
        LightGBMRegressor_f969f7c475b5-numLeaves: 6
}, {
        LightGBMRegressor_f969f7c475b5-maxDepth: 3,
        LightGBMRegressor_f969f7c475b5-numIterations: 10,
        LightGBMRegressor_f969f7c475b5-numLeaves: 6
}, {
        LightGBMRegressor_f969f7c475b5-maxDepth: 4,
        LightGBMRegressor_f969f7c475b5-numIterations: 10,
        LightGBMRegressor_f969f7c475b5-numLeaves: 6
}, {
        LightGBMRegressor_f969f7c475b5-maxDepth: 2,
        LightGBMRegressor_f969f7c475b5-numIterations: 10,
        LightGBMRegressor_f969f7c475b5-numLeaves: 9
}, {
        LightGBMRegressor_f969f7c475b5-maxDepth: 3,
        LightGBMRegressor_f969f7c475b5-numIterations: 10,
        LightGBMRegressor_f969f7c475b5-numLeaves: 9
}, {
        Lig...

import org.apache.spark.ml.evaluation.RegressionEvaluator

val evaluator = new RegressionEvaluator()
                .setLabelCol("price")
                .setPredictionCol("prediction")
                .setMetricName("rmse")

import org.apache.spark.ml.tuning.CrossValidator

val cv = new CrossValidator()
         .setEstimator(lightgbm)
         .setEvaluator(evaluator)
         .setEstimatorParamMaps(paramGrid)
         .setNumFolds(3)
```

```
val model = cv.fit(trainingData)

val predictions = model.transform(testData)

predictions.select("features","price","prediction").show
+--------------------+-----------------+------------------+
|            features|            price|        prediction|
+--------------------+-----------------+------------------+
|[17796.6311895433...|302355.83597895555| 965317.3181705693|
|[35454.7146594754...| 1077805.577726322|1093159.8506664087|
|[35608.9862370775...| 449331.5835333807|1061505.7131801855|
|[38868.2503114142...| 759044.6879907805|1061505.7131801855|
|[40752.7142433209...| 560598.5384309639| 974582.8481703462|
|[41007.4586732745...| 494742.5435776913| 881891.5646432829|
|[41533.0129597444...| 682200.3005599922| 966417.0064436384|
|[42258.7745410484...| 852703.2636757497|1070641.7611960804|
|[42940.1389392421...| 680418.7240122693|1028986.6314725328|
|[43192.1144092488...|1054606.9845532854|1087808.2361520242|
|[43241.9824225005...| 629657.6132544072| 889012.3734817103|
|[44328.2562966742...| 601007.3511604669| 828175.3829271109|
|[45347.1506816944...| 541953.9056802422| 860754.7467075661|
|[45546.6434075757...|   923830.33486809| 950407.7970842035|
|[45610.9384142094...|  961354.287727855|1175429.1179985087|
|[45685.2499205068...| 867714.3838490517|  828812.007346283|
|[45990.1237417814...|1043968.3994445396|1204501.1530193759|
|[46062.7542664558...| 675919.6815570832| 973273.6042265462|
|[46367.2058588838...|268050.81474351394| 761576.9192149616|
|[47467.4239151893...| 762144.9261238109| 951908.0117790927|
+--------------------+-----------------+------------------+
only showing top 20 rows

val rmse = evaluator.evaluate(predictions)

rmse: Double = 198601.74726198777
```

Let's extract the feature importance score for each feature.

```
val model = lightgbm.fit(trainingData)
```

```
model.getFeatureImportances("gain")
res7: Array[Double] = Array(1.110789482705408E15, 5.69355224816896E14,
3.25231517467648E14, 1.16104381056E13, 4.84685311277056E14)
```

By matching the order of the output in the list with the order of the feature in the feature vector (avg_area_income, avg_area_house_age, avg_area_num_rooms, avg_area_num_bedrooms, area_population), it looks like avg_area_income is our most important feature, followed by avg_area_house_age, area_population, and avg_area_num_rooms. The least important feature is avg_area_num_bedrooms.

Summary

I discussed some of the most popular supervised learning algorithms included in Spark MLlib as well as newer ones available externally such as XGBoost and LightGBM. While there is an abundance of documentation available online for XGBoost and LightGBM for Python, information and examples for Spark are limited. This chapter aims to help bridge the gap.

I refer you to `https://xgboost.readthedocs.io/en/latest` to learn more about XGBoost. For LightGBM, `https://lightgbm.readthedocs.io/en/latest` has the latest information. For a more in-depth coverage of the theory and math behind the classification and regression algorithms included in Spark MLlib, I refer you to *An Introduction to Statistical Learning* by Gareth James, Daniela Witten, Trevor Hastie, and Robert Tibshirani (Springer, 2017) and *The Elements of Statistical Learning by* Trevor Hastie, Robert Tibshirani, and Jerome Friedman (Springer, 2016). For more information on Spark MLlib, consult Apache Spark's *Machine Learning Library (MLlib) Guide* online: *https://spark.apache.org/docs/latest/ml-guide.html*.

References

i. Judea Pearl; "E PUR SI MUOVE (AND YET IT MOVES)," 2018, The Book Of Why: The New Science of Cause and Effect

ii. Apache Spark; "Multinomial logistic regression," spark.apache.org, 2019, https://spark.apache.org/docs/latest/ml-classification-regression.html#multinomial-logistic-regression

iii. Georgios Drakos; "Support Vector Machine vs Logistic Regression," towardsdatascience.com, 2018, https://towardsdatascience.com/support-vector-machine-vs-logistic-regression-94cc2975433f

iv. Apache Spark; "Multilayer perceptron classifier," spark.apache.org, 2019, https://spark.apache.org/docs/latest/ml-classification-regression.html#multilayer-perceptron-classifier

v. Analytics Vidhya Content Team; "A Complete Tutorial on Tree Based Modeling from Scratch (in R & Python)," AnalyticsVidhya.com, 2016, www.analyticsvidhya.com/blog/2016/04/complete-tutorial-tree-based-modeling-scratch-in-python/#one

vi. LightGBM; "Optimal Split for Categorical Features," lightgbm.readthedocs.io, 2019, https://lightgbm.readthedocs.io/en/latest/Features.html

vii. Joseph Bradley and Manish Amde; "Random Forests and Boosting in MLlib," Databricks, 2015, https://databricks.com/blog/2015/01/21/random-forests-and-boosting-in-mllib.html

viii. Analytics Vidhya Content Team; "An End-to-End Guide to Understand the Math behind XGBoost," analyticsvidhya.com, 2018, www.analyticsvidhya.com/blog/2018/09/an-end-to-end-guide-to-understand-the-math-behind-xgboost/

ix. Ben Gorman; "A Kaggle Master Explains Gradient Boosting," Kaggle.com, 2017, http://blog.kaggle.com/2017/01/23/a-kaggle-master-explains-gradient-boosting/

x. XGBoost; "Introduction to Boosted Trees," xgboost.readthedocs.io, 2019, https://xgboost.readthedocs.io/en/latest/tutorials/model.html

xi. Apache Spark; "Ensembles – RDD-based API," spark.apache.org, 2019, https://spark.apache.org/docs/latest/mllib-ensembles.html#gradient-boosted-trees-gbts

xii. Tianqi Chen; "When would one use Random Forests over Gradient Boosted Machines (GBMs)?," quora.com, 2015, www.quora.com/When-would-one-use-Random-Forests-over-Gradient-Boosted-Machines-GBMs

xiii. Aidan O'Brien, et. al.; "VariantSpark Machine Learning for Genomics Variants," CSIRO, 2018, https://bioinformatics.csiro.au/variantspark

xiv. Denis C. Bauer, et. al.; "Breaking the curse of dimensionality in Genomics using wide Random Forests," Databricks, 2017, https://databricks.com/blog/2017/07/26/breaking-the-curse-of-dimensionality-in-genomics-using-wide-random-forests.html

xv. Alessandro Lulli et. al.; "ReForeSt," github.com, 2017, https://github.com/alessandrolulli/reforest

xvi. Reforest; "How to learn a random forest classification model with ReForeSt," sites.google.com, 2019, https://sites.google.com/view/reforest/example?authuser=0

xvii. Apache Spark; "Ensembles - RDD-based API," spark.apache.org, 2019, https://spark.apache.org/docs/latest/mllib-ensembles.html#random-forests

xviii. CallMiner; "New research finds not valuing customers leads to $136 billion switching epidemic," CallMiner, 2018, www.globenewswire.com/news-release/2018/09/27/1577343/0/en/New-research-finds-not-valuing-customers-leads-to-136-billion-switching-epidemic.html

xix. Red Reichheld; "Prescription for cutting costs," Bain & Company, 2016, www2.bain.com/Images/BB_Prescription_cutting_costs.pdf

xx. Alex Lawrence; "Five Customer Retention Tips for Entrepreneurs," Forbes, 2012, www.forbes.com/sites/alexlawrence/2012/11/01/five-customer-retention-tips-for-entrepreneurs/

xxi. David Becks; "Churn in Telecom dataset," Kaggle, 2017, www.kaggle.com/becksddf/churn-in-telecoms-dataset

xxii. Jeffrey Shmain; "How-to Predict Telco Churn With Apache Spark MLlib," DZone, 2016, https://dzone.com/articles/how-to-predict-telco-churn-with-apache-spark-mllib

xxiii. Jake Hoare; "How is Variable Importance Calculated for a Random Forest," DisplayR, 2018, www.displayr.com/how-is-variable-importance-calculated-for-a-random-forest/

xxiv. Didrik Nielsen; "Tree Boosting With XGBoost," Norwegian University of Science and Technology, 2016, https://brage.bibsys.no/xmlui/bitstream/handle/11250/2433761/16128_FULLTEXT.pdf

xxv. Reena Shaw; "XGBoost: A Concise Technical Overview," KDNuggets, 2017, www.kdnuggets.com/2017/10/xgboost-concise-technical-overview.html

xxvi. Philip Hyunsu Cho; "Fast Histogram Optimized Grower, 8x to 10x Speedup," DMLC, 2017, https://github.com/dmlc/xgboost/issues/1950

xxvii. XGBoost; "Build an ML Application with XGBoost4J-Spark," xgboost.readthedocs.io, 2019, https://xgboost.readthedocs.io/en/latest/jvm/xgboost4j_spark_tutorial.html#pipeline-with-hyper-parameter-tunning

xxviii. Jason Brownlee; "How to Tune the Number and Size of Decision Trees with XGBoost in Python," machinelearningmastery.com, 2016, https://machinelearningmastery.com/tune-number-size-decision-trees-xgboost-python/

xxix. Laurae; "Benchmarking LightGBM: how fast is LightGBM vs xgboost?", medium.com, 2017, https://medium.com/implodinggradients/benchmarking-lightgbm-how-fast-is-lightgbm-vs-xgboost-15d224568031

xxx. LightGBM; "Optimization in Speed and Memory Usage," lightgbm.readthedocs.io, 2019, https://lightgbm.readthedocs.io/en/latest/Features.html

xxxi. David Marx; "Decision trees: leaf-wise (best-first) and level-wise tree traverse," stackexchange.com, 2018, https://datascience.stackexchange.com/questions/26699/decision-trees-leaf-wise-best-first-and-level-wise-tree-traverse

xxxii. LightGBM; "LightGBM Features," lightgbm.readthedocs.io, 2019, https://lightgbm.readthedocs.io/en/latest/Features.html

xxxiii. Szilard Pafka; "Performance of various open source GBM implementations," github.com, 2019, https://github.com/szilard/GBM-perf

xxxiv. Julio Antonio Soto; "What is the feature importance returned by 'gain' ?", github.com, 2018, https://github.com/Microsoft/LightGBM/issues/1842

xxxv. scikit-learn; "Linear Regression Example," scikit-learn.org, 2019, https://scikit-learn.org/stable/auto_examples/linear_model/plot_ols.html

xxxvi. Hongjian Li, et. al.; "Substituting random forest for multiple linear regression improves binding affinity prediction of scoring functions: Cyscore as a case study," nih.gov, 2014, `www.ncbi.nlm.nih.gov/pmc/articles/PMC4153907/`

xxxvii. Aariyan Panchal; "USA Housing.csv," Kaggle, 2018, `www.kaggle.com/aariyan101/usa-housingcsv`

xxxviii. Datasciencecentral; "RMSE: Root Mean Square Error," Datasciencecentral.com, 2016, `www.statisticshowto.datasciencecentral.com/rmse/`

CHAPTER 4

Unsupervised Learning

> *New knowledge is the most valuable commodity on earth. The more truth we have to work with, the richer we become.*
>
> —Kurt Vonnegut[i]

Unsupervised learning is a machine learning task that finds hidden patterns and structure in the dataset without the aid of labeled responses. Unsupervised learning is ideal when you only have access to input data and training data is unavailable or hard to obtain. Common methods include clustering, topic modeling, anomaly detection, and principal component analysis.

Clustering with K-Means

Clustering is an unsupervised machine learning task for grouping unlabeled observations that have some similarities. Popular clustering use cases include customer segmentation, fraud analysis, and anomaly detection. Clustering is also often used to generate training data for classifiers in cases where training data is scarce or unavailable. K-Means is one of the most popular unsupervised learning algorithms for clustering. Spark MLlib includes a more scalable implementation of K-means known as K-means||. Figure 4-1 shows K-means grouping the observations in the Iris dataset into three distinct clusters.

B. Quinto, *Next-Generation Machine Learning with Spark*, https://doi.org/10.1007/978-1-4842-5669-5_4

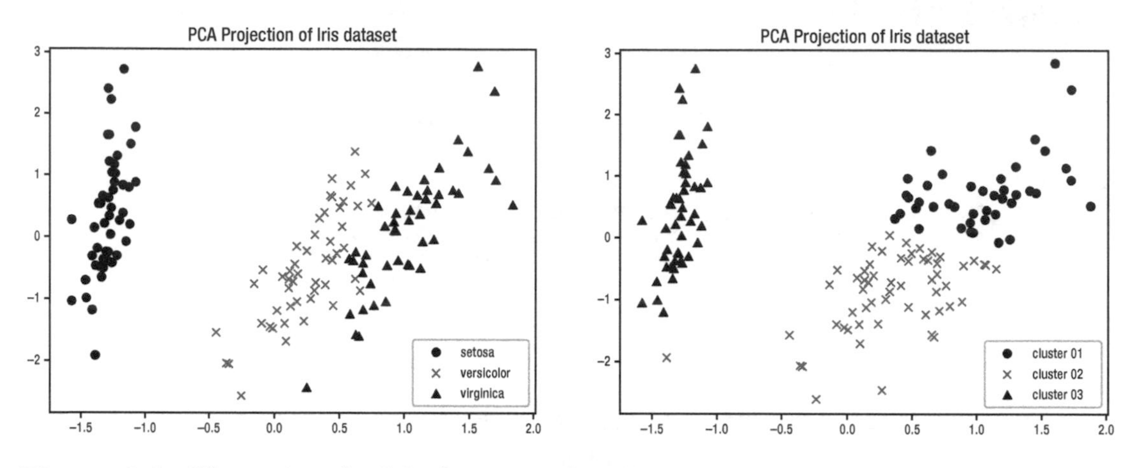

***Figure 4-1.** Clustering the Iris dataset using K-means*

Figure 4-2 shows the K-means algorithm in action. Observations are shown as squares, and cluster centroids are shown as triangles. Figure 4-2 (a) shows the original dataset. K-Means works by randomly assigning centroids that are used as the starting point for each cluster (Figure 4-2 (b) and (c)). The algorithm iteratively assigns each data point to the nearest centroid based on the Euclidean distance. It then calculates a new centroid for each cluster by calculating the mean of all the points that are part of that cluster (Figure 4-2 (d) and (e)). The algorithm stops iterating when a predefined number of iteration is reached or every data point is assigned to its nearest centroid and there are no more reassignments that can be performed (Figure 4-2 (f)).

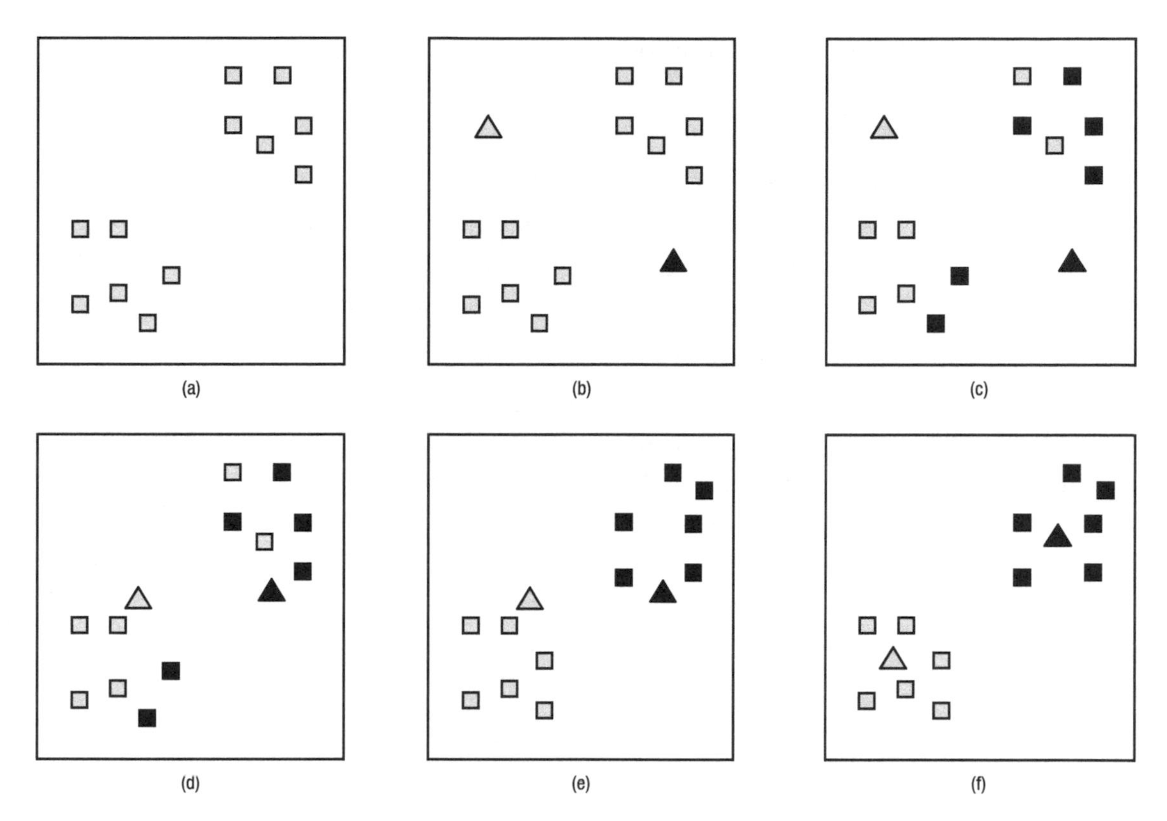

Figure 4-2. *K-Means algorithm in action*[ii]

K-Means requires the users to provide the number of clusters *k* to the algorithm. There are ways to find the optimal number of clusters for your dataset. We will discuss the elbow and silhouette method later in the chapter.

Example

Let's work on a simple customer segmentation example. We'll use a small dataset with seven observations and a mix of three categorical and two continuous features. Before we begin, we need to address another limitation of K-means. K-Means cannot directly work with categorical features such as gender *("M","F")*, marital status *("M","S")*, and state *("CA","NY")* and requires that all features be continuous. However, real-world datasets often include a combination of categorical and continuous features. Lucky for us, we can still use K-means with categorical features by converting them into numeric format.

This is not as straightforward as it sounds. For example, to convert marital status from its string representation *"M"* and "S" to a number, you may think that mapping 0 to *"M"* and 1 to *"S"* is appropriate for K-means. As you learned in Chapter 2, this is known as Integer or Label Encoding. But this introduces another wrinkle. Integers have a natural ordering *(0 < 1 < 2)* that some machine learning algorithms such as K-means may misinterpret, thinking that one categorical value is "greater" than the other simply because it is encoded as an integer, when in fact no such ordinal relationship exists in the data. This can produce unexpected results. To address this problem, we will use another type of encoding called one-hot encoding.[iii]

After converting the categorical feature to integer (using StringIndexer), we use one-hot encoding (using OneHotEncoderEstimator) to represent categorical features as binary vectors. For example, the state feature *("CA", "NY", "MA", "AZ")* is one-hot encoded in Table 4-1.

Table 4-1. *One-Hot Encoding State Feature*

CA	NY	MA	AZ
1	0	0	0
0	1	0	0
0	0	1	0
0	0	0	1

Feature scaling is another important preprocessing step for K-means. As discussed in Chapter 2, feature scaling is considered the best practice and a requirement for many machine learning algorithms that involve distance calculation. Feature scaling is especially important if the data is measured in different scales. Certain features may have significantly wide range of values, causing them to dominate other features. Feature scaling ensures that each feature proportionally contributes to the final distance. For our example, we will use the StandardScaler estimator to rescale our features to have a mean of 0 and unit variance (standard deviation of 1) as shown in Listing 4-1.

Listing 4-1. A Customer Segmentation Example Using K-Means

```
// Let's start with our example by creating some sample data.

val custDF = Seq(
(100, 29000,"M","F","CA",25),
(101, 36000,"M","M","CA",46),
(102, 5000,"S","F","NY",18),
(103, 68000,"S","M","AZ",39),
(104, 2000,"S","F","CA",16),
(105, 75000,"S","F","CA",41),
(106, 90000,"M","M","MA",47),
(107, 87000,"S","M","NY",38)
).toDF("customerid", "income","maritalstatus","gender","state","age")

// Perform some preprocessing steps.

import org.apache.spark.ml.feature.StringIndexer

val genderIndexer = new StringIndexer()
                    .setInputCol("gender")
                    .setOutputCol("gender_idx")

val stateIndexer = new StringIndexer()
                   .setInputCol("state")
                   .setOutputCol("state_idx")

val mstatusIndexer = new StringIndexer()
                     .setInputCol("maritalstatus")
                     .setOutputCol("maritalstatus_idx")

import org.apache.spark.ml.feature.OneHotEncoderEstimator

val encoder = new OneHotEncoderEstimator()
              .setInputCols(Array("gender_idx","state_idx",
              "maritalstatus_idx"))
              .setOutputCols(Array("gender_enc","state_enc",
              "maritalstatus_enc"))
```

```
val custDF2 = genderIndexer.fit(custDF).transform(custDF)

val custDF3 = stateIndexer.fit(custDF2).transform(custDF2)

val custDF4 = mstatusIndexer.fit(custDF3).transform(custDF3)

custDF4.select("gender_idx","state_idx","maritalstatus_idx").show

+----------+---------+-----------------+
|gender_idx|state_idx|maritalstatus_idx|
+----------+---------+-----------------+
|       0.0|      0.0|              1.0|
|       1.0|      0.0|              1.0|
|       0.0|      1.0|              0.0|
|       1.0|      3.0|              0.0|
|       0.0|      0.0|              0.0|
|       0.0|      0.0|              0.0|
|       1.0|      2.0|              1.0|
|       1.0|      1.0|              0.0|
+----------+---------+-----------------+

val custDF5 = encoder.fit(custDF4).transform(custDF4)

custDF5.printSchema

root
 |-- customerid: integer (nullable = false)
 |-- income: integer (nullable = false)
 |-- maritalstatus: string (nullable = true)
 |-- gender: string (nullable = true)
 |-- state: string (nullable = true)
 |-- age: integer (nullable = false)
 |-- gender_idx: double (nullable = false)
 |-- state_idx: double (nullable = false)
 |-- maritalstatus_idx: double (nullable = false)
 |-- gender_enc: vector (nullable = true)
 |-- state_enc: vector (nullable = true)
 |-- maritalstatus_enc: vector (nullable = true)

custDF5.select("gender_enc","state_enc","maritalstatus_enc").show
```

```
+-------------+------------+----------------+
|   gender_enc|   state_enc|maritalstatus_enc|
+-------------+------------+----------------+
|(1,[0],[1.0])|(3,[0],[1.0])|       (1,[],[])|
|    (1,[],[])|(3,[0],[1.0])|       (1,[],[])|
|(1,[0],[1.0])|(3,[1],[1.0])|   (1,[0],[1.0])|
|    (1,[],[])|    (3,[],[])|   (1,[0],[1.0])|
|(1,[0],[1.0])|(3,[0],[1.0])|   (1,[0],[1.0])|
|(1,[0],[1.0])|(3,[0],[1.0])|   (1,[0],[1.0])|
|    (1,[],[])|(3,[2],[1.0])|       (1,[],[])|
|    (1,[],[])|(3,[1],[1.0])|   (1,[0],[1.0])|
+-------------+------------+----------------+

import org.apache.spark.ml.feature.VectorAssembler

val assembler = new VectorAssembler()
                .setInputCols(Array("income","gender_enc", "state_enc",
                "maritalstatus_enc", "age"))
                .setOutputCol("features")

val custDF6 = assembler.transform(custDF5)

custDF6.printSchema

root
 |-- customerid: integer (nullable = false)
 |-- income: integer (nullable = false)
 |-- maritalstatus: string (nullable = true)
 |-- gender: string (nullable = true)
 |-- state: string (nullable = true)
 |-- age: integer (nullable = false)
 |-- gender_idx: double (nullable = false)
 |-- state_idx: double (nullable = false)
 |-- maritalstatus_idx: double (nullable = false)
 |-- gender_enc: vector (nullable = true)
 |-- state_enc: vector (nullable = true)
 |-- maritalstatus_enc: vector (nullable = true)
 |-- features: vector (nullable = true)
```

```
custDF6.select("features").show(false)

+---------------------------------+
|features                         |
+---------------------------------+
|[29000.0,1.0,1.0,0.0,0.0,0.0,25.0]|
|(7,[0,2,6],[36000.0,1.0,46.0])    |
|[5000.0,1.0,0.0,1.0,0.0,1.0,18.0] |
|(7,[0,5,6],[68000.0,1.0,39.0])    |
|[2000.0,1.0,1.0,0.0,0.0,1.0,16.0] |
|[75000.0,1.0,1.0,0.0,0.0,1.0,41.0]|
|(7,[0,4,6],[90000.0,1.0,47.0])    |
|[87000.0,0.0,0.0,1.0,0.0,1.0,38.0]|
+---------------------------------+

import org.apache.spark.ml.feature.StandardScaler

val scaler = new StandardScaler()
             .setInputCol("features")
             .setOutputCol("scaledFeatures")
             .setWithStd(true)
             .setWithMean(false)

val custDF7 = scaler.fit(custDF6).transform(custDF6)

custDF7.printSchema

root
 |-- customerid: integer (nullable = false)
 |-- income: integer (nullable = false)
 |-- maritalstatus: string (nullable = true)
 |-- gender: string (nullable = true)
 |-- state: string (nullable = true)
 |-- age: integer (nullable = false)
 |-- gender_idx: double (nullable = false)
 |-- state_idx: double (nullable = false)
 |-- maritalstatus_idx: double (nullable = false)
 |-- gender_enc: vector (nullable = true)
```

```
 |-- state_enc: vector (nullable = true)
 |-- maritalstatus_enc: vector (nullable = true)
 |-- features: vector (nullable = true)
 |-- scaledFeatures: vector (nullable = true)

custDF7.select("scaledFeatures").show(8,65)

+-----------------------------------------------------------------+
| scaledFeatures                                                  |
+-----------------------------------------------------------------+
|[0.8144011366375091,1.8708286933869707,1.8708286933869707,0.0,...|
|(7,[0,2,6],[1.0109807213431148,1.8708286933869707,3.7319696616...|
|[0.140413989075432б,1.8708286933869707,0.0,2.160246899469287,0...|
|(7,[0,5,6],[1.9096302514258834,1.9321835661585918,3.1640612348...|
|[0.05616559563017304,1.8708286933869707,1.8708286933869707,0.0...|
|[2.106209836131489,1.8708286933869707,1.8708286933869707,0.0,0...|
|(7,[0,4,6],[2.5274518033577866,2.82842712474619,3.813099436871...|
|[2.443203409912527,0.0,0.0,2.160246899469287,0.0,1.93218356615...|
+-----------------------------------------------------------------+

// We'll create two clusters.

import org.apache.spark.ml.clustering.KMeans

val kmeans = new KMeans()
             .setFeaturesCol("scaledFeatures")
             .setPredictionCol("prediction")
             .setK(2)

import org.apache.spark.ml.Pipeline

val pipeline = new Pipeline()
               .setStages(Array(genderIndexer, stateIndexer,
               mstatusIndexer, encoder, assembler, scaler, kmeans))

val model = pipeline.fit(custDF)

val clusters = model.transform(custDF)
```

```
clusters.select("customerid","income","maritalstatus",
               "gender","state","age","prediction")
               .show

+----------+------+-------------+------+-----+---+----------+
|customerid|income|maritalstatus|gender|state|age|prediction|
+----------+------+-------------+------+-----+---+----------+
|       100| 29000|            M|     F|   CA| 25|         1|
|       101| 36000|            M|     M|   CA| 46|         0|
|       102|  5000|            S|     F|   NY| 18|         1|
|       103| 68000|            S|     M|   AZ| 39|         0|
|       104|  2000|            S|     F|   CA| 16|         1|
|       105| 75000|            S|     F|   CA| 41|         0|
|       106| 90000|            M|     M|   MA| 47|         0|
|       107| 87000|            S|     M|   NY| 38|         0|
+----------+------+-------------+------+-----+---+----------+

import org.apache.spark.ml.clustering.KMeansModel

val model = pipeline.stages.last.asInstanceOf[KMeansModel]

model.clusterCenters.foreach(println)
[1.9994952044341603,0.37416573867739417,0.7483314773547883,0.4320493798938574,
0.565685424949238,1.159310139695155,3.4236765156588613]
[0.3369935737810382,1.8708286933869707,1.247219128924647,0.7200822998230956,
0.0,1.288122377439061,1.5955522466340666]
```

We evaluate our cluster by computing Within Set Sum of Squared Errors (WSSSE). Examining the WSSSE using the “elbow method” is often used to assist in determining the optimal number of clusters. The elbow method works by fitting the model with a range of values for k and plotting it against the WSSSE. Visually examine the line chart, and if it resembles a flexed arm, the point where it bends on the curve (the “elbow”) indicates the most optimal value for k.

```
val wssse = model.computeCost(custDF)
wssse: Double = 32.09801038868844
```

Another way to evaluate cluster quality is by computing the silhouette coefficient score. The silhouette score provides a metric of how close each point in one cluster is to points in the other clusters. The larger the silhouette score, the better the quality of the cluster. A score closer to 1 indicates that the points are closer to the centroid of the cluster. A score closer to 0 indicates that the points are closer to other clusters, and a negative value indicates the points may have been designated to the wrong cluster.

```
import org.apache.spark.ml.evaluation.ClusteringEvaluator

val evaluator = new ClusteringEvaluator()

val silhouette = evaluator.evaluate(clusters)

silhouette: Double = 0.6722088068201866
```

Topic Modeling with Latent Dirichlet Allocation (LDA)

Latent Dirichlet Allocation (LDA) was developed in 2003 by David M. Blei, Andrew Ng, and Michael Jordan, although a similar algorithm used in population genetics was also proposed by Jonathan K. Pritchard, Matthew Stephens, and Peter Donnelly in 2000. LDA, as applied to machine learning, is based on a graphical model and is the first algorithm included in Spark MLlib built on GraphX. Latent Dirichlet Allocation is widely used for topic modeling. Topic models automatically derive the themes (or topics) in a group of documents (Figure 4-3). These topics can be used for content-based recommendations, document classification, dimensionality reduction, and featurization.

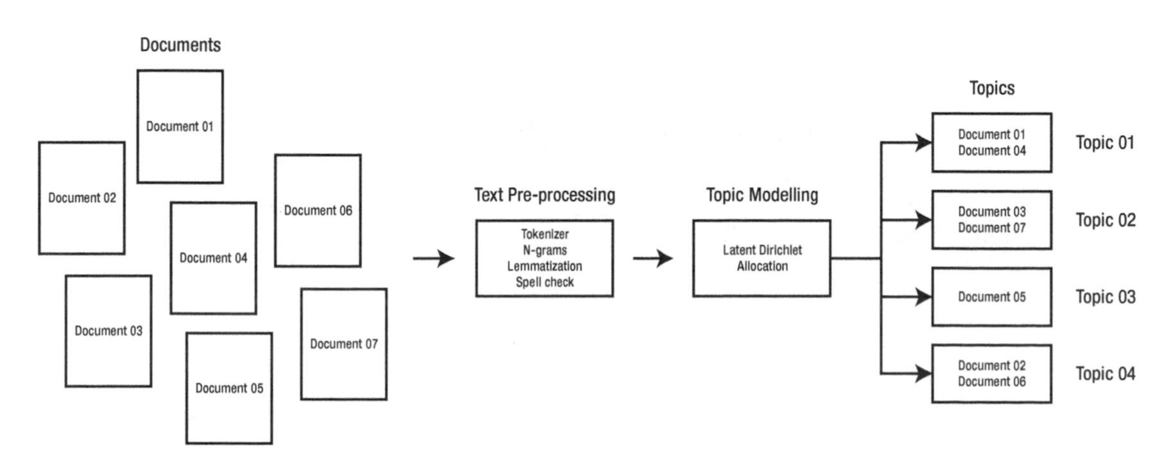

Figure 4-3. *Grouping documents by topic using Latent Dirichlet Allocation*

Despite Spark MLlib's extensive text mining and preprocessing capabilities, it lacks several features found in most enterprise-grade NLP libraries[iv] such as lemmatization, stemming, and sentiment analysis, to mention a few. We will need some of these features for our topic modeling example later in the chapter. This is a good time to introduce Stanford CoreNLP for Spark and Spark NLP from John Snow Labs.

Stanford CoreNLP for Spark

Stanford CoreNLP is a professional-grade NLP library developed by the NLP research group at Stanford University. CoreNLP supports multiple languages such as Arabic, Chinese, English, French, and German.[v] It provides a native Java API, as well as a web API and command line interface. There are also third-party APIs for major programming languages such as R, Python, Ruby, and Lua. Xiangrui Meng, a software engineer from Databricks, developed the Stanford CoreNLP wrapper for Spark (see Listing 4-2).

Listing 4-2. A Brief Introduction to Stanford CoreNLP for Spark

```
spark-shell --packages databricks:spark-corenlp:0.4.0-spark2.4-scala2.11
--jars stanford-corenlp-3.9.1-models.jar

import spark.implicits._
import org.apache.spark.sql.types._

val dataDF = Seq(
(1, "Kevin Durant was the 2019 All-Star NBA Most Valuable Player."),
(2, "Stephen Curry is the best clutch three-point shooter in the NBA."),
(3, "My game is not as good as it was 20 years ago."),
(4, "Michael Jordan is the greatest NBA player of all time."),
(5, "The Lakers currently have one of the worst performances in the NBA."))
.toDF("id", "text")

dataDF.show(false)
```

```
+---+-----------------------------------------------------------+
|id |text                                                       |
+---+-----------------------------------------------------------+
|1  |Kevin Durant was the 2019 All-Star NBA Most Valuable Player.   |
|2  |Stephen Curry is the best clutch three-point shooter in the NBA. |
|3  |My game is not as good as it was 20 years ago.             |
|4  |Michael Jordan is the greatest NBA player of all time.     |
|5  |The Lakers currently have one of the worst performances in the NBA.|
+---+-----------------------------------------------------------+

// Stanford CoreNLP lets you chain text processing functions. Let's split
// the document into sentences and then tokenize the sentences into words.

import com.databricks.spark.corenlp.functions._

val dataDF2 = dataDF
              .select(explode(ssplit('text)).as('sen))
              .select('sen, tokenize('sen).as('words))

dataDF2.show(5,30)

+------------------------------+------------------------------+
|                           sen|                         words|
+------------------------------+------------------------------+
|Kevin Durant was the 2019 A...|[Kevin, Durant, was, the, 2...|
|Stephen Curry is the best c...|[Stephen, Curry, is, the, b...|
|My game is not as good as i...|[My, game, is, not, as, goo...|
|Michael Jordan is the great...|[Michael, Jordan, is, the, ...|
|The Lakers currently have o...|[The, Lakers, currently, ha...|
+------------------------------+------------------------------+

// Perform sentiment analysis on the sentences. The scale
// ranges from 0 for strong negative to 4 for strong positive.

val dataDF3 = dataDF
              .select(explode(ssplit('text)).as('sen))
              .select('sen, tokenize('sen).as('words), sentiment('sen).
              as('sentiment))
```

```
dataDF3.show(5,30)

+------------------------------+------------------------------+---------+
|                           sen|                         words|sentiment|
+------------------------------+------------------------------+---------+
|Kevin Durant was the 2019 A...|[Kevin, Durant, was, the, 2...|        1|
|Stephen Curry is the best c...|[Stephen, Curry, is, the, b...|        3|
|My game is not as good as i...|[My, game, is, not, as, goo...|        1|
|Michael Jordan is the great...|[Michael, Jordan, is, the, ...|        3|
|The Lakers currently have o...|[The, Lakers, currently, ha...|        1|
+------------------------------+------------------------------+---------+
```

Visit Databricks' CoreNLP GitHub page for a complete list of features available in Stanford CoreNLP for Spark.

Spark NLP from John Snow Labs

The Spark NLP library from John Snow Labs natively supports the Spark ML Pipeline API. It is written in Scala and includes both Scala and Python APIs. It includes several advanced features such as a tokenizer, lemmatizer, stemmer, entity, and date extractors, part of speech tagger, sentence boundary detection, spellchecker, and named entity recognition, to mention a few.

Annotators provide the NLP capabilities in Spark NLP. An annotation is the result of a Spark NLP operation. There are two types of annotators, *Annotator Approaches and Annotator Model. Annotator Approaches* represent Spark MLlib estimators. It fits a model with data to produce an annotator model or transformer. An *Annotator Model* is a transformer that takes a dataset and adds a column with the result of the annotation. Because they are represented as Spark estimators and transformers, annotators can be easily integrated with the Spark Pipeline API. Spark NLP provides several ways for users to access its capabilities.[vi]

Pre-trained Pipelines

Spark NLP includes pre-trained pipelines for quick text annotations. Spark NLP offers a pre-trained pipeline named explain_document_ml that accepts text as input (see Listing 4-3). The pre-trained pipeline contains popular text processing capabilities and provides a quick and dirty way to use Spark NLP without too much hassle.

Listing 4-3. Spark NLP Pre-trained Pipeline Example

```
spark-shell --packages JohnSnowLabs:spark-nlp:2.1.0

import com.johnsnowlabs.nlp.pretrained.PretrainedPipeline

val annotations = PretrainedPipeline("explain_document_ml").annotate
("I visited Greece last summer. It was a great trip. I went swimming in
Mykonos.")

annotations("sentence")
res7: Seq[String] = List(I visited Greece last summer.,
It was a great trip., I went swimming in Mykonos.)

annotations("token")
res8: Seq[String] = List(I, visited, Greece, last, summer, .,
It, was, a, great, trip, ., I, went, swimming, in, Mykonos, .)

annotations("lemma")
res9: Seq[String] = List(I, visit, Greece, last, summer, .,
It, be, a, great, trip, ., I, go, swim, in, Mykonos, .)
```

Pre-trained Pipelines with Spark DataFrames

The pre-trained pipelines also work with Spark DataFrames, as shown in Listing 4-4.

Listing 4-4. Spark NLP Pre-trained Pipelines with Spark DataFrames

```
val data = Seq("I visited Greece last summer. It was a great trip. I went
swimming in Mykonos.").toDF("text")

val annotations = PretrainedPipeline("explain_document_ml").transform(data)

annotations.show()

+--------------------+--------------------+--------------------+
|                text|            document|            sentence|
+--------------------+--------------------+--------------------+
|I visited Greece ...|[[document, 0, 77...|[[document, 0, 28...|
+--------------------+--------------------+--------------------+
```

```
+--------------------+
|               token|
+--------------------+
|[[token, 0, 0, I,...|
+--------------------+
+--------------------+--------------------+--------------------+
|             checked|               lemma|                stem|
+--------------------+--------------------+--------------------+
|[[token, 0, 0, I,...|[[token, 0, 0, I,...|[[token, 0, 0, i,...|
+--------------------+--------------------+--------------------+

+--------------------+
|                 pos|
+--------------------+
|[[pos, 0, 0, PRP,...|
+--------------------+
```

Pre-trained Pipelines with Spark MLlib Pipelines

You can use the pre-trained pipelines with Spark MLlib pipelines (see Listing 4-5). Note that a special transformer called *Finisher* is required to display tokens in a human-readable format.

Listing 4-5. Pre-trained Pipelines with Spark MLlib Pipeline Example

```
import com.johnsnowlabs.nlp.Finisher
import org.apache.spark.ml.Pipeline

val data = Seq("I visited Greece last summer. It was a great trip. I went
swimming in Mykonos.").toDF("text")

val finisher = new Finisher()
               .setInputCols("sentence", "token", "lemma")

val explainPipeline = PretrainedPipeline("explain_document_ml").model

val pipeline = new Pipeline()
               .setStages(Array(explainPipeline,finisher))
```

```
pipeline.fit(data).transform(data).show(false)

+--------------------------------------------+
|text                                        |
+--------------------------------------------+
|I visited Greece last summer. It was a great trip.|
+--------------------------------------------+

+----------------------------+
| text                       |
+----------------------------+
|I went swimming in Mykonos. |
+----------------------------+

+-----------------------------------------------------+
|finished_sentence                                    |
+-----------------------------------------------------+
|[I visited Greece last summer., It was a great trip. |
+-----------------------------------------------------+

+-----------------------------+
| finished_sentence           |
+-----------------------------+
|,I went swimming in Mykonos.]|
+-----------------------------+

+-----------------------------------------------------------------+
|finished_token                                                   |
+-----------------------------------------------------------------+
|[I, visited, Greece, last, summer, ., It, was, a, great, trip, .,|
+-----------------------------------------------------------------+

+--------------------------------------------------------------+
|finished_lemma                                                |
+--------------------------------------------------------------+
|[I, visit, Greece, last, summer, ., It, be, a, great, trip, .,|
+--------------------------------------------------------------+
```

```
+--------------------------------+
| finished_lemma                 |
+--------------------------------+
|, I, go, swim, in, Mykonos, .]  |
+--------------------------------+
```

Creating Your Own Spark MLlib Pipeline

You can use the annotators directly from your own Spark MLlib pipeline, as shown in Listing 4-6.

Listing 4-6. Creating Your Own Spark MLlib Pipeline Example

```
import com.johnsnowlabs.nlp.base._
import com.johnsnowlabs.nlp.annotator._
import org.apache.spark.ml.Pipeline

val data = Seq("I visited Greece last summer. It was a great trip. I went
swimming in Mykonos.").toDF("text")

val documentAssembler = new DocumentAssembler()
                        .setInputCol("text")
                        .setOutputCol("document")

val sentenceDetector = new SentenceDetector()
                       .setInputCols(Array("document"))
                       .setOutputCol("sentence")

val regexTokenizer = new Tokenizer()
                     .setInputCols(Array("sentence"))
                     .setOutputCol("token")

val finisher = new Finisher()
               .setInputCols("token")
               .setCleanAnnotations(false)

val pipeline = new Pipeline()
               .setStages(Array(documentAssembler,
               sentenceDetector,regexTokenizer,finisher))
```

```
pipeline.fit(Seq.empty[String].toDF("text"))
        .transform(data)
        .show()

+--------------------+--------------------+--------------------+
|                text|            document|            sentence|
+--------------------+--------------------+--------------------+
|I visited Greece ...|[[document, 0, 77...|[[document, 0, 28...|
+--------------------+--------------------+--------------------+

+--------------------+--------------------+
|               token|      finished_token|
+--------------------+--------------------+
|[[token, 0, 0, I,...|[I, visited, Gree...|
+--------------------+--------------------+
```

Spark NLP LightPipeline

Spark NLP provides another class of pipeline called LightPipeline. It is similar to Spark MLlib pipelines, but instead of taking advantage of Spark's distributed processing capabilities, execution happens locally. LightPipeline is appropriate when processing small amounts of data and low-latency execution is required (see Listing 4-7).

Listing 4-7. Spark NLP LightPipelines Example

```
import com.johnsnowlabs.nlp.base._

val trainedModel = pipeline.fit(Seq.empty[String].toDF("text"))

val lightPipeline = new LightPipeline(trainedModel)

lightPipeline.annotate("I visited Greece last summer.")
```

Spark NLP OCR Module

Spark NLP includes an OCR module that lets users create Spark DataFrames from PDF files. The OCR module is not included in the core Spark NLP library. To use it you need to include a separate package and specify an additional repository as you can see in my spark-shell command in Listing 4-8.

Listing 4-8. Spark NLP OCR Module Example

```
spark-shell --packages JohnSnowLabs:spark-nlp:2.1.0,com.johnsnowlabs.
nlp:spark-nlp-ocr_2.11:2.1.0,javax.media.jai:com.springsource.javax.media.
jai.core:1.1.3
      --repositories http://repo.spring.io/plugins-release

import com.johnsnowlabs.nlp.util.io.OcrHelper

val myOcrHelper = new OcrHelper

val data = myOcrHelper.createDataset(spark, "/my_pdf_files/")

val documentAssembler = new DocumentAssembler().setInputCol("text")

documentAssembler.transform(data).select("text","filename").show(1,45)

+-------------------------------------------+
|                                       text|
+-------------------------------------------+
|this is a PDF document. Have a great day.  |
+-------------------------------------------+

+---------------------------------------------+
|                                    filename |
+---------------------------------------------+
|file:/my_pdf_files/document.pdf              |
+---------------------------------------------+
```

Spark NLP is a powerful library that includes a lot more features not covered in this introduction. To learn more about Spark NLP, visit `http://nlp.johnsnowlabs.com`.

Example

We can proceed with our topic modeling example now that we have everything we need. We will use Latent Dirichlet Allocation to classify over one million news headlines by topic published over a period of 15 years. The dataset, which can be downloaded from Kaggle, was provided by the Australian Broadcasting Corporation and made available by Rohit Kulkarni.

We can use either the Spark NLP from John Snow Labs or the Stanford CoreNLP package to provide us with additional text processing capabilities. We'll use the Stanford CoreNLP package for this example (see Listing 4-9).

Listing 4-9. Topic Modeling with LDA

```
spark-shell --packages databricks:spark-corenlp:0.4.0-spark2.4-scala2.11
--jars stanford-corenlp-3.9.1-models.jar

import org.apache.spark.sql.functions._
import org.apache.spark.sql.types._
import org.apache.spark.sql._

// Define the schema.

var newsSchema = StructType(Array (
StructField("publish_date",   IntegerType, true),
StructField("headline_text",   StringType, true)
    ))

// Read the data.

val dataDF = spark.read.format("csv")
             .option("header", "true")
             .schema(newsSchema)
             .load("abcnews-date-text.csv")

// Inspect the data.

dataDF.show(false)
+------------+--------------------------------------------------+
|publish_date|headline_text                                     |
+------------+--------------------------------------------------+
|20030219    |aba decides against community broadcasting licence|
|20030219    |act fire witnesses must be aware of defamation    |
|20030219    |a g calls for infrastructure protection summit    |
|20030219    |air nz staff in aust strike for pay rise          |
|20030219    |air nz strike to affect australian travellers     |
|20030219    |ambitious olsson wins triple jump                 |
|20030219    |antic delighted with record breaking barca        |
|20030219    |aussie qualifier stosur wastes four memphis match |
```

```
|20030219    |aust addresses un security council over iraq      |
|20030219    |australia is locked into war timetable opp        |
|20030219    |australia to contribute 10 million in aid to iraq |
|20030219    |barca take record as robson celebrates birthday in|
|20030219    |bathhouse plans move ahead                        |
|20030219    |big hopes for launceston cycling championship     |
|20030219    |big plan to boost paroo water supplies            |
|20030219    |blizzard buries united states in bills            |
|20030219    |brigadier dismisses reports troops harassed in    |
|20030219    |british combat troops arriving daily in kuwait    |
|20030219    |bryant leads lakers to double overtime win        |
|20030219    |bushfire victims urged to see centrelink          |
+------------+--------------------------------------------------+
only showing top 20 rows

// Remove punctuations.

val dataDF2 = dataDF
             .withColumn("headline_text",
             regexp_replace((dataDF("headline_text")), "[^a-zA-Z0-9 ]", ""))

// We will use Stanford CoreNLP to perform lemmatization. As discussed
// earlier, lemmatization derives the root form of inflected words. For
// example, "camping", "camps", "camper", and "camped" are all inflected
// forms of "camp". Reducing inflected words to its root form helps reduce
// the complexity of performing natural language processing. A similar
// process known as stemming also reduces inflected words to their root
// form, but it does so by crudely chopping off affixes, even though the
// root form may not be a valid word. In contrast, lemmatization ensures
// that the inflected words are reduced to a valid root word through the
// morphological analysis of words and the use of a vocabulary.[vii]

import com.databricks.spark.corenlp.functions._

val dataDF3 = dataDF2

.select(explode(ssplit('headline_text)).as('sen))
             .select('sen, lemma('sen)
             .as('words))
```

```
dataDF3.show
+--------------------+--------------------+
|                 sen|               words|
+--------------------+--------------------+
|aba decides again...|[aba, decide, aga...|
|act fire witnesse...|[act, fire, witne...|
|a g calls for inf...|[a, g, call, for,...|
|air nz staff in a...|[air, nz, staff, ...|
|air nz strike to ...|[air, nz, strike,...|
|ambitious olsson ...|[ambitious, olsso...|
|antic delighted w...|[antic, delighted...|
|aussie qualifier ...|[aussie, qualifie...|
|aust addresses un...|[aust, address, u...|
|australia is lock...|[australia, be, l...|
|australia to cont...|[australia, to, c...|
|barca take record...|[barca, take, rec...|
|bathhouse plans m...|[bathhouse, plan,...|
|big hopes for lau...|[big, hope, for, ...|
|big plan to boost...|[big, plan, to, b...|
|blizzard buries u...|[blizzard, bury, ...|
|brigadier dismiss...|[brigadier, dismi...|
|british combat tr...|[british, combat,...|
|bryant leads lake...|[bryant, lead, la...|
|bushfire victims ...|[bushfire, victim...|
+--------------------+--------------------+
only showing top 20 rows

// We'll remove stop words such as "a", "be", and "to". Stop
// words have no contribution to the meaning of a document.

import org.apache.spark.ml.feature.StopWordsRemover

val remover = new StopWordsRemover()
              .setInputCol("words")
              .setOutputCol("filtered_stopwords")
```

```
val dataDF4 = remover.transform(dataDF3)

dataDF4.show

+--------------------+--------------------+--------------------+
|                 sen|               words|  filtered_stopwords|
+--------------------+--------------------+--------------------+
|aba decides again...|[aba, decide, aga...|[aba, decide, com...|
|act fire witnesse...|[act, fire, witne...|[act, fire, witne...|
|a g calls for inf...|[a, g, call, for,...|[g, call, infrast...|
|air nz staff in a...|[air, nz, staff, ...|[air, nz, staff, ...|
|air nz strike to ...|[air, nz, strike,...|[air, nz, strike,...|
|ambitious olsson ...|[ambitious, olsso...|[ambitious, olsso...|
|antic delighted w...|[antic, delighted...|[antic, delighted...|
|aussie qualifier ...|[aussie, qualifie...|[aussie, qualifie...|
|aust addresses un...|[aust, address, u...|[aust, address, u...|
|australia is lock...|[australia, be, l...|[australia, lock,...|
|australia to cont...|[australia, to, c...|[australia, contr...|
|barca take record...|[barca, take, rec...|[barca, take, rec...|
|bathhouse plans m...|[bathhouse, plan,...|[bathhouse, plan,...|
|big hopes for lau...|[big, hope, for, ...|[big, hope, launc...|
|big plan to boost...|[big, plan, to, b...|[big, plan, boost...|
|blizzard buries u...|[blizzard, bury, ...|[blizzard, bury, ...|
|brigadier dismiss...|[brigadier, dismi...|[brigadier, dismi...|
|british combat tr...|[british, combat,...|[british, combat,...|
|bryant leads lake...|[bryant, lead, la...|[bryant, lead, la...|
|bushfire victims ...|[bushfire, victim...|[bushfire, victim...|
+--------------------+--------------------+--------------------+
only showing top 20 rows

// Generate n-grams. n-grams are a sequence of "n" number of words often
// used to discover the relationship of words in a document.  For example,
// "Los Angeles" is a bigram. "Los" and "Angeles" are unigrams. "Los" and
// "Angeles" when considered as individual units may not mean much, but it
// is more meaningful when combined as a single entity "Los Angeles".
// Determining the optimal number of "n" is dependent on the use case
// and the language used in the document.[viii] For our example, we'll
generate // a unigram, bigram, and trigram.
```

```
import org.apache.spark.ml.feature.NGram

val unigram = new NGram()
              .setN(1)
              .setInputCol("filtered_stopwords")
              .setOutputCol("unigram_words")

val dataDF5 = unigram.transform(dataDF4)

dataDF5.printSchema
root
 |-- sen: string (nullable = true)
 |-- words: array (nullable = true)
 |    |-- element: string (containsNull = true)
 |-- filtered_stopwords: array (nullable = true)
 |    |-- element: string (containsNull = true)
 |-- unigram_words: array (nullable = true)
 |    |-- element: string (containsNull = false)

val bigram = new NGram()
             .setN(2)
             .setInputCol("filtered_stopwords")
             .setOutputCol("bigram_words")

val dataDF6 = bigram.transform(dataDF5)

dataDF6.printSchema
root
 |-- sen: string (nullable = true)
 |-- words: array (nullable = true)
 |    |-- element: string (containsNull = true)
 |-- filtered_stopwords: array (nullable = true)
 |    |-- element: string (containsNull = true)
 |-- unigram_words: array (nullable = true)
 |    |-- element: string (containsNull = false)
 |-- bigram_words: array (nullable = true)
 |    |-- element: string (containsNull = false)
```

```
val trigram = new NGram()
              .setN(3)
              .setInputCol("filtered_stopwords")
              .setOutputCol("trigram_words")

val dataDF7 = trigram.transform(dataDF6)

dataDF7.printSchema
root
 |-- sen: string (nullable = true)
 |-- words: array (nullable = true)
 |    |-- element: string (containsNull = true)
 |-- filtered_stopwords: array (nullable = true)
 |    |-- element: string (containsNull = true)
 |-- unigram_words: array (nullable = true)
 |    |-- element: string (containsNull = false)
 |-- bigram_words: array (nullable = true)
 |    |-- element: string (containsNull = false)
 |-- trigram_words: array (nullable = true)
 |    |-- element: string (containsNull = false)

// We combine the unigram, bigram, and trigram into a single vocabulary.
// We will concatenate and store the words in a column "ngram_words"
// using Spark SQL.

dataDF7.createOrReplaceTempView("dataDF7")

val dataDF8 = spark.sql("select sen,words,filtered_stopwords,unigram_words,
bigram_words,trigram_words,concat(concat(unigram_words,bigram_words),
trigram_words) as ngram_words from dataDF7")

dataDF8.printSchema
root
 |-- sen: string (nullable = true)
 |-- words: array (nullable = true)
 |    |-- element: string (containsNull = true)
 |-- filtered_stopwords: array (nullable = true)
 |    |-- element: string (containsNull = true)
```

```
 |-- unigram_words: array (nullable = true)
 |    |-- element: string (containsNull = false)
 |-- bigram_words: array (nullable = true)
 |    |-- element: string (containsNull = false)
 |-- trigram_words: array (nullable = true)
 |    |-- element: string (containsNull = false)
 |-- ngram_words: array (nullable = true)
 |    |-- element: string (containsNull = false)

dataDF8.select("ngram_words").show(20,65)
+-----------------------------------------------------------------+
|ngram_words                                                      |
+-----------------------------------------------------------------+
|[aba, decide, community, broadcasting, licence, aba decide, de...|
|[act, fire, witness, must, aware, defamation, act fire, fire w...|
|[g, call, infrastructure, protection, summit, g call, call inf...|
|[air, nz, staff, aust, strike, pay, rise, air nz, nz staff, st...|
|[air, nz, strike, affect, australian, traveller, air nz, nz st...|
|[ambitious, olsson, win, triple, jump, ambitious olsson, olsso...|
|[antic, delighted, record, break, barca, antic delighted, deli...|
|[aussie, qualifier, stosur, waste, four, memphis, match, aussi...|
|[aust, address, un, security, council, iraq, aust address, add...|
|[australia, lock, war, timetable, opp, australia lock, lock wa...|
|[australia, contribute, 10, million, aid, iraq, australia cont...|
|[barca, take, record, robson, celebrate, birthday, barca take,...|
|[bathhouse, plan, move, ahead, bathhouse plan, plan move, move...|
|[big, hope, launceston, cycling, championship, big hope, hope ...|
|[big, plan, boost, paroo, water, supplies, big plan, plan boos...|
|[blizzard, bury, united, state, bill, blizzard bury, bury unit...|
|[brigadier, dismiss, report, troops, harass, brigadier dismiss...|
|[british, combat, troops, arrive, daily, kuwait, british comba...|
|[bryant, lead, laker, double, overtime, win, bryant lead, lead...|
|[bushfire, victim, urge, see, centrelink, bushfire victim, vic...|
+-----------------------------------------------------------------+
only showing top 20 rows
```

```
// Use CountVectorizer to convert the text data to vectors of token counts.

import org.apache.spark.ml.feature.{CountVectorizer, CountVectorizerModel}

val cv = new CountVectorizer()
         .setInputCol("ngram_words")
         .setOutputCol("features")

val cvModel = cv.fit(dataDF8)

val dataDF9 = cvModel.transform(dataDF8)

val vocab = cvModel.vocabulary

vocab: Array[String] = Array(police, man, new, say, plan, charge, call,
council, govt, fire, court, win, interview, back, kill, australia, find,
death, urge, face, crash, nsw, report, water, get, australian, qld, take,
woman, wa, attack, sydney, year, change, murder, hit, health, jail, claim,
day, child, miss, hospital, car, home, sa, help, open, rise, warn, school,
world, market, cut, set, accuse, die, seek, drug, make, boost, may, coast,
government, ban, job, group, fear, mp, two, talk, service, farmer,
minister, election, fund, south, road, continue, lead, worker, first,
national, test, arrest, work, rural, go, power, price, cup, final, concern,
green, china, mine, fight, labor, trial, return, flood, deal, north, case,
push, pm, melbourne, law, driver, one, nt, want, centre, record, ...

// We use IDF to scale the features generated by CountVectorizer.
// Scaling features generally improves performance.

import org.apache.spark.ml.feature.IDF

val idf = new IDF()
          .setInputCol("features")
          .setOutputCol("features2")

val idfModel = idf.fit(dataDF9)

val dataDF10 = idfModel.transform(dataDF9)
```

```
dataDF10.select("features2").show(20,65)
+-----------------------------------------------------------------+
| features2                                                       |
+-----------------------------------------------------------------+
|(262144,[154,1054,1140,15338,19285],[5.276861439995834,6.84427...|
|(262144,[9,122,711,727,3141,5096,23449],[4.189486226673463,5.1...|
|(262144,[6,734,1165,1177,1324,43291,96869],[4.070620900306447,...|
|(262144,[48,121,176,208,321,376,424,2183,6231,12147,248053],[4...|
|(262144,[25,176,208,376,764,3849,12147,41079,94670,106284],[4....|
|(262144,[11,1008,1743,10833,128493,136885],[4.2101466208496285...|
|(262144,[113,221,3099,6140,9450,16643],[5.120230688038215,5.54...|
|(262144,[160,259,483,633,1618,4208,17750,187744],[5.3211036079...|
|(262144,[7,145,234,273,321,789,6163,10334,11101,32988],[4.0815...|
|(262144,[15,223,1510,5062,5556],[4.393970862600795,5.555011224...|
|(262144,[15,145,263,372,541,3896,15922,74174,197210],[4.393970...|
|(262144,[27,113,554,1519,3099,13499,41664,92259],[4.5216508634...|
|(262144,[4,131,232,5636,6840,11444,37265],[3.963488754657374,5...|
|(262144,[119,181,1288,1697,2114,49447,80829,139670],[5.1266204...|
|(262144,[4,23,60,181,2637,8975,9664,27571,27886],[3.9634887546...|
|(262144,[151,267,2349,3989,7631,11862],[5.2717309555002725,5.6...|
|(262144,[22,513,777,12670,33787,49626],[4.477068652869369,6.16...|
|(262144,[502,513,752,2211,5812,7154,30415,104812],[6.143079025...|
|(262144,[11,79,443,8222,8709,11447,194715],[4.2101466208496285...|
|(262144,[18,146,226,315,2877,5160,19389,42259],[4.414350240692...|
+-----------------------------------------------------------------+
only showing top 20 rows

// The scaled features could then be passed to LDA.

import org.apache.spark.ml.clustering.LDA

val lda = new LDA()
          .setK(30)
          .setMaxIter(10)

val model = lda.fit(dataDF10)
```

```
val topics = model.describeTopics

topics.show(20,30)
+-----+------------------------------+------------------------------+
|topic|                   termIndices|                   termWeights|
+-----+------------------------------+------------------------------+
|    0|[2, 7, 16, 9482, 9348, 5, 1...|[1.817876125380732E-4, 1.09...|
|    1|[974, 2, 3, 5189, 5846, 541...|[1.949552388785536E-4, 1.89...|
|    2|[2253, 4886, 12, 6767, 3039...|[2.7922272919208327E-4, 2.4...|
|    3|[6218, 6313, 5762, 3387, 27...|[1.6618313204146235E-4, 1.6...|
|    4|[0, 1, 39, 14, 13, 11, 2, 1...|[1.981809243111437E-4, 1.22...|
|    5|[4, 7, 22, 11, 2, 3, 79, 92...|[2.49620962563534E-4, 2.032...|
|    6|[15, 32, 319, 45, 342, 121,...|[2.885684164769467E-5, 2.45...|
|    7|[2298, 239, 1202, 3867, 431...|[3.435238376348344E-4, 3.30...|
|    8|[0, 4, 110, 3, 175, 38, 8, ...|[1.0177738516279581E-4, 8.7...|
|    9|[1, 19, 10, 2, 7, 8, 5, 0, ...|[2.2854683602607976E-4, 1.4...|
|   10|[1951, 1964, 16, 33, 1, 5, ...|[1.959705576881449E-4, 1.92...|
|   11|[12, 89, 72, 3, 92, 63, 62,...|[4.167255720848278E-5, 3.19...|
|   12|[4, 23, 13, 22, 73, 18, 70,...|[1.1641833113477034E-4, 1.1...|
|   13|[12, 1, 5, 16, 185, 132, 24...|[0.008769073702733892, 0.00...|
|   14|[9151, 13237, 3140, 14, 166...|[8.201099412213086E-5, 7.85...|
|   15|[9, 1, 0, 11, 3, 15, 32, 52...|[0.0032039727688580703, 0.0...|
|   16|[1, 10, 5, 56, 27, 3, 16, 1...|[5.252120584885086E-5, 4.05...|
|   17|[12, 1437, 4119, 1230, 5303...|[5.532790361864421E-4, 2.97...|
|   18|[12, 2459, 7836, 8853, 7162...|[6.862552774818539E-4, 1.83...|
|   19|[21, 374, 532, 550, 72, 773...|[0.0024665346250921432, 0.0...|
+-----+------------------------------+------------------------------+
only showing top 20 rows

// Determine the max size of the vocabulary.

model.vocabSize
res27: Int = 262144

// Extract the topic words. The describeTopics method returns the
// dictionary indices from CountVectorizer's output. We will use a custom
// user-defined function to map the words to the indices.[ix]
```

```
import scala.collection.mutable.WrappeddArray
import org.apache.spark.sql.functions.udf

val extractWords = udf( (x : WrappedArray[Int]) => { x.map(i => vocab(i)) })

val topics = model
            .describeTopics
            .withColumn("words", extractWords(col("termIndices")))

topics.select("topic","termIndices","words").show(20,30)
+-----+------------------------------+------------------------------+
|topic|                   termIndices|                         words|
+-----+------------------------------+------------------------------+
|    0|[2, 7, 16, 9482, 9348, 5, 1...|[new, council, find, abuse ...|
|    1|[974, 2, 3, 5189, 5846, 541...|[2016, new, say, china sea,...|
|    2|[2253, 4886, 12, 6767, 3039...|[nathan, interview nathan, ...|
|    3|[6218, 6313, 5762, 3387, 27...|[new guinea, papua new guin...|
|    4|[0, 1, 39, 14, 13, 11, 2, 1...|[police, man, day, kill, ba...|
|    5|[4, 7, 22, 11, 2, 3, 79, 92...|[plan, council, report, win...|
|    6|[15, 32, 319, 45, 342, 121,...|[australia, year, india, sa...|
|    7|[2298, 239, 1202, 3867, 431...|[sach, tour, de, tour de, d...|
|    8|[0, 4, 110, 3, 175, 38, 8, ...|[police, plan, nt, say, fun...|
|    9|[1, 19, 10, 2, 7, 8, 5, 0, ...|[man, face, court, new, cou...|
|   10|[1951, 1964, 16, 33, 1, 5, ...|[vic country, vic country h...|
|   11|[12, 89, 72, 3, 92, 63, 62,...|[interview, price, farmer, ...|
|   12|[4, 23, 13, 22, 73, 18, 70,...|[plan, water, back, report,...|
|   13|[12, 1, 5, 16, 185, 132, 24...|[interview, man, charge, fi...|
|   14|[9151, 13237, 3140, 14, 166...|[campese, interview terry, ...|
|   15|[9, 1, 0, 11, 3, 15, 32, 52...|[fire, man, police, win, sa...|
|   16|[1, 10, 5, 56, 27, 3, 16, 1...|[man, court, charge, die, t...|
|   17|[12, 1437, 4119, 1230, 5303...|[interview, redback, 666, s...|
|   18|[12, 2459, 7836, 8853, 7162...|[interview, simon, intervie...|
|   19|[21, 374, 532, 550, 72, 773...|[nsw, asylum, seeker, asylu...|
+-----+------------------------------+------------------------------+
only showing top 20 rows
```

```
// Extract the term weights from describeTopics.

val wordsWeight = udf( (x : WrappedArray[Int],
y : WrappedArray[Double]) =>
{ x.map(i => vocab(i)).zip(y)}
)

val topics2 = model
             .describeTopics
             .withColumn("words", wordsWeight(col("termIndices"),
             col("termWeights")))

val topics3 = topics2
             .select("topic", "words")
             .withColumn("words", explode(col("words")))
topics3.show(50,false)
+-----+------------------------------------------------+
|topic|words                                           |
+-----+------------------------------------------------+
|0    |[new, 1.4723785654465323E-4]                    |
|0    |[council, 1.242876719889358E-4]                 |
|0    |[thursday, 1.1710009304019913E-4]               |
|0    |[grandstand thursday, 1.0958369194828903E-4]    |
|0    |[two, 8.119593156862581E-5]                     |
|0    |[charge, 7.321024120305904E-5]                  |
|0    |[find, 6.98723717903146E-5]                     |
|0    |[burley griffin, 6.474176573486395E-5]          |
|0    |[claim, 6.448801852215021E-5]                   |
|0    |[burley, 6.390953777977556E-5]                  |
|1    |[say, 1.9595383103126804E-4]                    |
|1    |[new, 1.7986957579978078E-4]                    |
|1    |[murder, 1.7156446166835784E-4]                 |
|1    |[las, 1.6793241095301546E-4]                    |
|1    |[vegas, 1.6622904053495525E-4]                  |
|1    |[las vegas, 1.627321199362179E-4]               |
|1    |[2016, 1.4906599207615762E-4]                   |
|1    |[man, 1.3653760511354596E-4]                    |
```

```
|1    |[call, 1.3277357539424398E-4]                       |
|1    |[trump, 1.250570735309821E-4]                       |
|2    |[ntch, 5.213678388314454E-4]                        |
|2    |[ntch podcast, 4.6907569870744537E-4]               |
|2    |[podcast, 4.625754070258578E-4]                     |
|2    |[interview, 1.2297477650126824E-4]                  |
|2    |[trent, 9.319817855283612E-5]                       |
|2    |[interview trent, 8.967384560094343E-5]             |
|2    |[trent robinson, 7.256857525120274E-5]              |
|2    |[robinson, 6.888930961680287E-5]                    |
|2    |[interview trent robinson, 6.821800839623336E-5]|
|2    |[miss, 6.267572268770148E-5]                        |
|3    |[new, 8.244153432249302E-5]                         |
|3    |[health, 5.269269109549137E-5]                      |
|3    |[change, 5.1481361386635024E-5]                     |
|3    |[first, 3.474601129571304E-5]                       |
|3    |[south, 3.335342687995096E-5]                       |
|3    |[rise, 3.3245575277669534E-5]                       |
|3    |[country, 3.26422466284622E-5]                      |
|3    |[abuse, 3.25594250748893E-5]                        |
|3    |[start, 3.139959761950907E-5]                       |
|3    |[minister, 3.1327427652213426E-5]                   |
|4    |[police, 1.756612187665565E-4]                      |
|4    |[man, 1.2903801461819285E-4]                        |
|4    |[petero, 8.259870531430337E-5]                      |
|4    |[kill, 8.251557569137285E-5]                        |
|4    |[accuse grant, 8.187325944352362E-5]                |
|4    |[accuse grant bail, 7.609807356711693E-5]           |
|4    |[find, 7.219731162848223E-5]                        |
|4    |[attack, 6.804063612991027E-5]                      |
|4    |[day, 6.772554893634948E-5]                         |
|4    |[jail, 6.470525327671485E-5]                        |
+-----+------------------------------------------------+
only showing top 50 rows
```

```
// Finally, we split the word and the weight into separate fields.

val topics4 = topics3
             .select(col("topic"), col("words")
             .getField("_1").as("word"), col("words")
             .getField("_2").as("weight"))

topics4.show(50, false)

+-----+------------------------+---------------------+
|topic|word                    |weight               |
+-----+------------------------+---------------------+
|0    |new                     |1.4723785654465323E-4|
|0    |council                 |1.242876719889358E-4 |
|0    |thursday                |1.1710009304019913E-4|
|0    |grandstand thursday     |1.0958369194828903E-4|
|0    |two                     |8.119593156862581E-5 |
|0    |charge                  |7.321024120305904E-5 |
|0    |find                    |6.98723717903146E-5  |
|0    |burley griffin          |6.474176573486395E-5 |
|0    |claim                   |6.448801852215021E-5 |
|0    |burley                  |6.390953777977556E-5 |
|1    |say                     |1.9595383103126804E-4|
|1    |new                     |1.7986957579978078E-4|
|1    |murder                  |1.7156446166835784E-4|
|1    |las                     |1.6793241095301546E-4|
|1    |vegas                   |1.6622904053495525E-4|
|1    |las vegas               |1.627321199362179E-4 |
|1    |2016                    |1.4906599207615762E-4|
|1    |man                     |1.3653760511354596E-4|
|1    |call                    |1.3277357539424398E-4|
|1    |trump                   |1.250570735309821E-4 |
|2    |ntch                    |5.213678388314454E-4 |
|2    |ntch podcast            |4.6907569870744537E-4|
|2    |podcast                 |4.625754070258578E-4 |
|2    |interview               |1.2297477650126824E-4|
|2    |trent                   |9.319817855283612E-5 |
```

```
|2    |interview trent         |8.967384560094343E-5 |
|2    |trent robinson          |7.256857525120274E-5 |
|2    |robinson                |6.888930961680287E-5 |
|2    |interview trent robinson|6.821800839623336E-5 |
|2    |miss                    |6.267572268770148E-5 |
|3    |new                     |8.244153432249302E-5 |
|3    |health                  |5.269269109549137E-5 |
|3    |change                  |5.1481361386635024E-5|
|3    |first                   |3.474601129571304E-5 |
|3    |south                   |3.335342687995096E-5 |
|3    |rise                    |3.3245575277669534E-5|
|3    |country                 |3.26422466284622E-5  |
|3    |abuse                   |3.25594250748893E-5  |
|3    |start                   |3.139959761950907E-5 |
|3    |minister                |3.1327427652213426E-5|
|4    |police                  |1.756612187665565E-4 |
|4    |man                     |1.2903801461819285E-4|
|4    |petero                  |8.259870531430337E-5 |
|4    |kill                    |8.251557569137285E-5 |
|4    |accuse grant            |8.187325944352362E-5 |
|4    |accuse grant bail       |7.609807356711693E-5 |
|4    |find                    |7.219731162848223E-5 |
|4    |attack                  |6.804063612991027E-5 |
|4    |day                     |6.772554893634948E-5 |
|4    |jail                    |6.470525327671485E-5 |
+-----+------------------------+---------------------+
only showing top 50 rows
```

I'm only showing the top 50 rows for brevity, displaying 4 topics out of 30. If you inspect the words in each topic closely, you'll see repeating themes that can be used to classify the headlines.

Anomaly Detection with Isolation Forest

Anomaly or outlier detection identifies rare observations that deviate significantly and stand out from majority of the dataset. It is frequently used in discovering fraudulent financial transactions, identifying cybersecurity threats, or performing predictive maintenance, to mention a few use cases. Anomaly detection is a popular research area in the field of machine learning. Several anomaly detection techniques have been invented throughout the years with varying degrees of effectiveness. For this chapter, I will cover one of the most effective anomaly detection techniques called Isolation Forest. Isolation Forest is a tree-based ensemble algorithm for anomaly detection that was developed by Fei Tony Liu, Kai Ming Ting, and Zhi-Hua Zhou.[x]

Unlike most anomaly detection techniques, Isolation Forest tries to explicitly detect actual outliers instead of identifying normal data points. Isolation Forest operates based on the fact that there are usually a small number of outliers in a dataset and are therefore prone to the process of isolation.[xi] Isolating outliers from normal data points is efficient since it requires fewer conditions. In contrast, isolating normal data points generally involves more conditions. As shown in Figure 4-4 (b), the anomalous data point was isolated with just one division, while the normal data points took five divisions to isolate. When data is represented as a tree structure, anomalies are more likely to be closer to the root node at a much shallower depth than normal data points. As shown in Figure 4-4 (a), the outlier (8, 12) has a tree depth of 1, while the normal data point (9, 15) has a tree depth of 5.

Isolation Forest does not require feature scaling since the distance threshold used in detecting outliers is based on tree depth. It works great for large and small datasets and does not require training datasets since it's an unsupervised learning technique.[xiii]

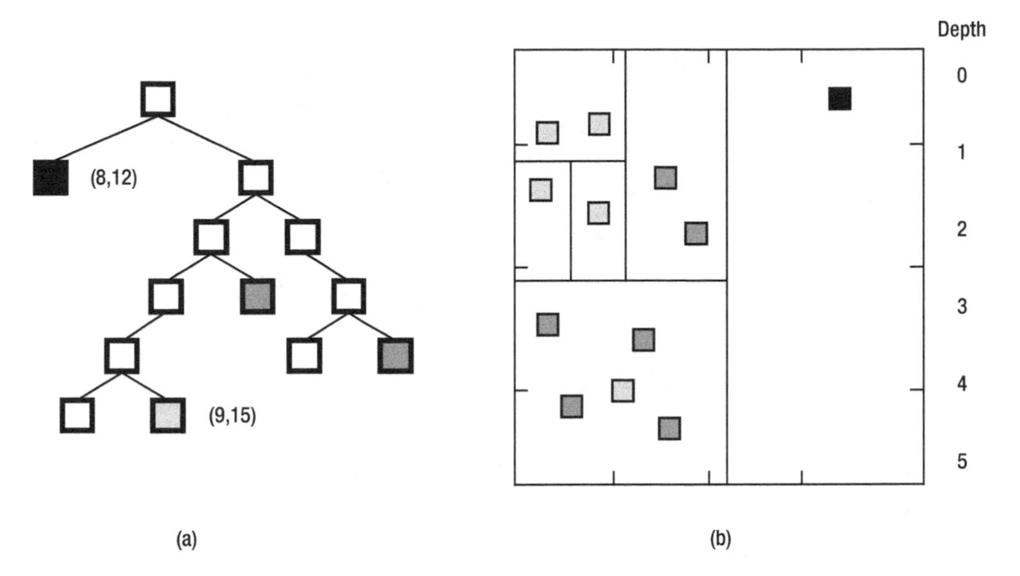

Figure 4-4. *Number of divisions required to isolate anomalous and normal data points*[xii] *with Isolation Forest*

Similar to other tree-based ensembles, Isolation Forest is built on a collection of decision trees known as *isolation trees,* with each tree having a subset of the entire dataset. An anomaly score is computed as the average anomaly score of the trees in the forest. The anomaly score is derived from the number of conditions required to split a data point. An anomaly score close to 1 signifies an anomaly, while a score less than 0.5 signifies non-anomalous observations (Figure 4-5).

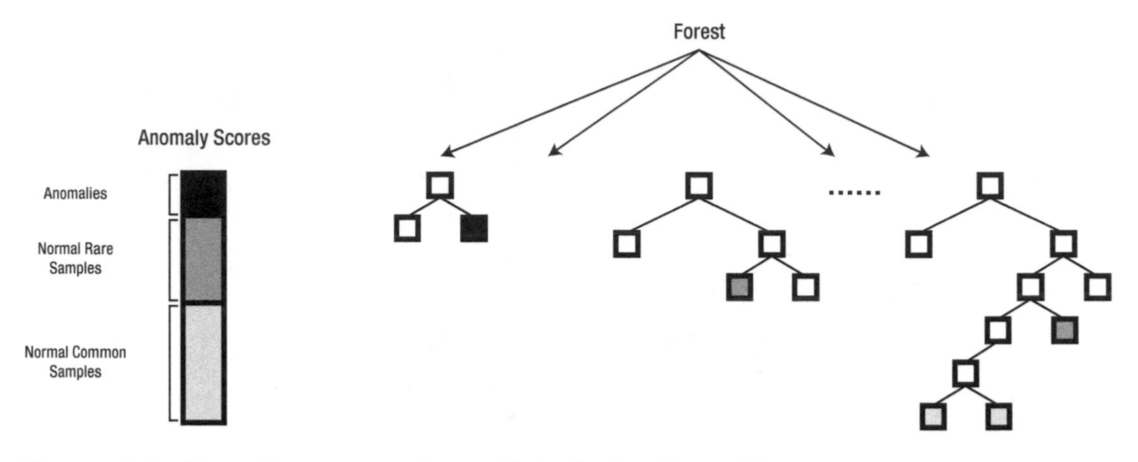

***Figure 4-5.** Detecting anomalies with Isolation Forest*[xiv]

Isolation Forest has outperformed other anomaly detection methods in both accuracy and performance. Figures 4-6 and 4-7 show a performance comparison of Isolation Forest against One-Class Support Vector Machine, another well-known outlier detection algorithm.[xv] The first test evaluated both algorithms against normal observations that belonged to a single group (Figure 4-6), while the second test evaluated both algorithms against observations that belonged to two uneven clusters (Figure 4-7). In both cases Isolation Forest performed better than One-Class Support Vector Machine.

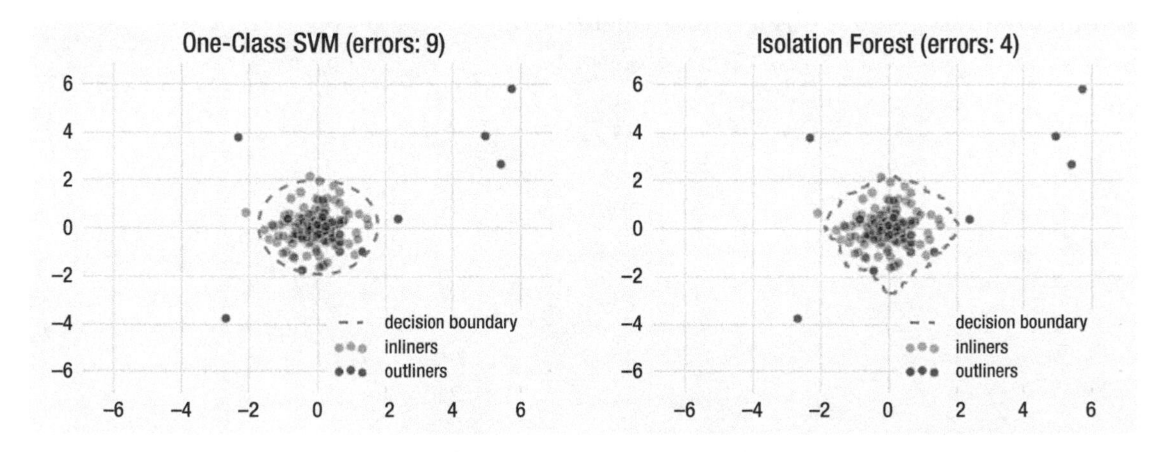

***Figure 4-6.** Isolation Forest vs. One-Class Support Vector Machine – normal observations, single group (image courtesy of Alejandro Correa Bahnsen)*

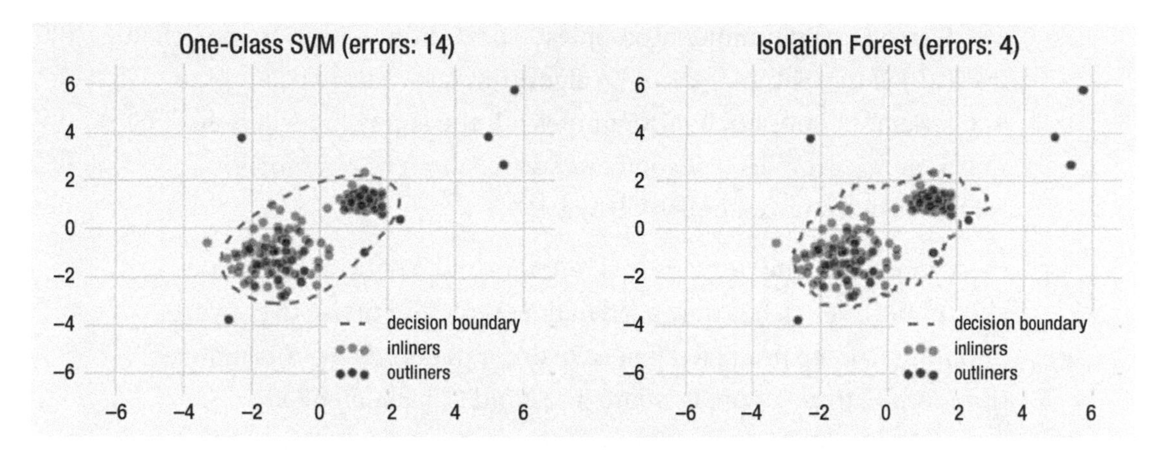

Figure 4-7. *Isolation Forest vs. One-Class Support Vector Machine – uneven clusters (image courtesy of Alejandro Correa Bahnsen)*

Spark-iForest is an implementation of the Isolation Forest algorithm in Spark developed by Fangzhou Yang with the help of Jie Fang and several contributors. It is available as an external third-party package and not included in the standard Apache Spark MLlib library. You can find more information about Spark-iForest as well as the latest JAR file by visiting the Spark-iForest GitHub page `https://github.com/titicaca/spark-iforest`.[xvi]

Parameters

This is a list of parameters supported by Spark-iForest. As you can see, some of the parameters are similar to other tree-based ensembles such as Random Forest.

- *maxFeatures*: The number of features to draw from data to train each tree (>0). If maxFeatures <= 1, the algorithm will draw maxFeatures ∗ totalFeatures features. If maxFeatures > 1, the algorithm will draw maxFeatures features.
- *maxDepth*: The height limit used in constructing a tree (>0). The default value will be about log2(numSamples).
- *numTrees*: The number of trees in the iforest model (>0).

- *maxSamples*: The number of samples to draw from data to train each tree (>0). If maxSamples <= 1, the algorithm will draw maxSamples ∗ totalSample samples. If maxSamples > 1, the algorithm will draw maxSamples samples. The total memory is about maxSamples ∗ numTrees ∗ 4 + maxSamples ∗ 8 bytes.
- *contamination*: The proportion of outliers in the dataset; the value should be in (0, 1). It is only used in the prediction phase to convert anomaly score to predicted labels. In order to enhance performance, our method to get anomaly score threshold is calculated by approxQuantile. You can set the param approxQuantileRelativeError greater than 0 in order to calculate an approximate quantile threshold of anomaly scores for large dataset.
- *approxQuantileRelativeError*: The relative error for approximate quantile calculation (0 <= value <= 1); default is 0 for calculating the exact value, which would be expensive for large datasets.
- *bootstrap*: If true, individual trees are fit on random subsets of the training data sampled with replacement. If false, sampling without replacement is performed.
- *seed*: The seed used by the random number generator.
- *featuresCol*: Features column name, default "features".
- *anomalyScoreCol*: Anomaly score column name, default “anomalyScore”.
- *predictionCol*: Prediction column name, default “prediction”.[xvii]

Example

We will use the Spark-iForest to predict the occurrence of breast cancer (Listing 4-10) using the Wisconsin Breast Cancer Dataset (Table 4-2), available from the UCI Machine Learning Repository.[xviii]

Table 4-2. *Wisconsin Breast Cancer Dataset*

Index	Feature	Domain
1	Sample code number	id number
2	Clump thickness	1–10
3	Uniformity of cell size	1–10
4	Uniformity of cell shape	1–10
5	Marginal adhesion	1–10
6	Single epithelial cell size	1–10
7	Bare nuclei	1–10
8	Bland chromatin	1–10
9	Normal nucleoli	1–10
10	Mitoses	1–10
11	Class	(2 for benign, 4 for malignant)

Listing 4-10. Anomaly Detection with Isolation Forest

```
spark-shell --jars spark-iforest-1.0-SNAPSHOT.jar

import org.apache.spark.sql.types._

var dataSchema = StructType(Array(
StructField("id", IntegerType, true),
StructField("clump_thickness", IntegerType, true),
StructField("ucell_size", IntegerType, true),
StructField("ucell_shape", IntegerType, true),
StructField("marginal_ad", IntegerType, true),
StructField("se_cellsize", IntegerType, true),
StructField("bare_nuclei", IntegerType, true),
StructField("bland_chromatin", IntegerType, true),
StructField("normal_nucleoli", IntegerType, true),
StructField("mitosis", IntegerType, true),
StructField("class", IntegerType, true)
    ))
```

```
val dataDF = spark.read.option("inferSchema", "true")
             .schema(dataSchema)
             .csv("/files/breast-cancer-wisconsin.csv")

dataDF.printSchema

//The dataset contain 16 rows with missing attribute values.
//We'll remove them for this exercise.

val dataDF2 = dataDF.filter("bare_nuclei is not null")

val seed = 1234

val Array(trainingData, testData) = dataDF2.randomSplit(Array(0.8, 0.2),
seed)

import org.apache.spark.ml.feature.StringIndexer

val labelIndexer = new StringIndexer().setInputCol("class").
setOutputCol("label")

import org.apache.spark.ml.feature.VectorAssembler

val assembler = new VectorAssembler()
                .setInputCols(Array("clump_thickness",
                "ucell_size", "ucell_shape", "marginal_ad", "se_cellsize",
"bare_nuclei", "bland_chromatin", "normal_nucleoli", "mitosis"))
                .setOutputCol("features")

import org.apache.spark.ml.iforest._

val iForest = new IForest()
              .setMaxSamples(150)
              .setContamination(0.30)
              .setBootstrap(false)
              .setSeed(seed)
              .setNumTrees(100)
              .setMaxDepth(50)

val pipeline = new Pipeline()
      .setStages(Array(labelIndexer, assembler, iForest))
```

```
val model = pipeline.fit(trainingData)

val predictions = model.transform(testData)

predictions.select("id","features","anomalyScore","prediction").show()

+------+--------------------+-------------------+----------+
|    id|            features|       anomalyScore|prediction|
+------+--------------------+-------------------+----------+
| 63375|[9.0,1.0,2.0,6.0,...| 0.6425205920636737|       1.0|
| 76389|[10.0,4.0,7.0,2.0...| 0.6475157383643779|       1.0|
| 95719|[6.0,10.0,10.0,10...| 0.6413247885878359|       1.0|
|242970|[5.0,7.0,7.0,1.0,...| 0.6156526231532693|       1.0|
|353098|[4.0,1.0,1.0,2.0,...|0.45686731187686386|       0.0|
|369565|[4.0,1.0,1.0,1.0,...|0.45957810648090186|       0.0|
|390840|[8.0,4.0,7.0,1.0,...| 0.6387497388682214|       1.0|
|412300|[10.0,4.0,5.0,4.0...| 0.6104797020175959|       1.0|
|466906|[1.0,1.0,1.0,1.0,...|0.41857428772927696|       0.0|
|476903|[10.0,5.0,7.0,3.0...| 0.6152957125696049|       1.0|
|486283|[3.0,1.0,1.0,1.0,...|0.47218763124223706|       0.0|
|557583|[5.0,10.0,10.0,10...| 0.6822227844447365|       1.0|
|636437|[1.0,1.0,1.0,1.0,...|0.41857428772927696|       0.0|
|654244|[1.0,1.0,1.0,1.0,...| 0.4163657637214968|       0.0|
|657753|[3.0,1.0,1.0,4.0,...|0.49314746153500594|       0.0|
|666090|[1.0,1.0,1.0,1.0,...|0.45842258207090547|       0.0|
|688033|[1.0,1.0,1.0,1.0,...|0.41857428772927696|       0.0|
|690557|[5.0,1.0,1.0,1.0,...| 0.4819098604217553|       0.0|
|704097|[1.0,1.0,1.0,1.0,...| 0.4163657637214968|       0.0|
|770066|[5.0,2.0,2.0,2.0,...| 0.5125093127301371|       0.0|
+------+--------------------+-------------------+----------+
only showing top 20 rows
```

We can't use BinaryClassificationEvaluator to evaluate the Isolation Forest model since it expects the raw Predictions field to be present in the output. Spark-iForest produces an anomalyScore field instead of rawPrediction. We will use BinaryClassificationMetrics instead to evaluate the model.

```
import org.apache.spark.mllib.evaluation.BinaryClassificationMetrics

val binaryMetrics = new BinaryClassificationMetrics(
predictions.select("prediction", "label").rdd.map {
case Row(prediction: Double, label: Double) => (prediction, label)
}
)

println(s"AUC: ${binaryMetrics.areaUnderROC()}")

AUC: 0.9532866199532866
```

Dimensionality Reduction with Principal Component Analysis

Principal component analysis (PCA) is an unsupervised machine learning technique used for reducing the dimensionality of the feature space. It detects correlations between features and generates a reduced number of linearly uncorrelated features while retaining most of the variance in the original dataset. These more compact, linearly uncorrelated features are called *principal components*. The principal components are sorted in descending order of their explained variance. Dimensionality reduction is essential when there are a high number of features in your dataset. Machine learning use cases in the fields of genomics and industrial analytics, for instance, usually involve thousands or even millions of features. High dimensionality makes models more complex, increasing the chances of overfitting. Adding more features at a certain point will actually decrease the performance of the model. Moreover, training on high-dimensional data requires significant computing resources. These are collectively known as the *curse of dimensionality*. Dimensionality reduction techniques aim to overcome the curse of dimensionality.

Note that the principal components generated by PCA will not be interpretable. This is a deal-breaker in situations where you need to understand why the prediction was made. Furthermore, it is essential to standardize your dataset before applying PCA to prevent features that are on the largest scale to be considered more important than other features.

Example

For our example, we will use PCA on the Iris dataset to project four-dimensional feature vectors into two-dimensional principal components (see Listing 4-11).

Listing 4-11. Reducing Dimensions with PCA

```
import org.apache.spark.ml.feature.{PCA, VectorAssembler}
import org.apache.spark.ml.feature.StringIndexer
import org.apache.spark.sql.types._

val irisSchema = StructType(Array (
StructField("sepal_length",   DoubleType, true),
StructField("sepal_width",   DoubleType, true),
StructField("petal_length",   DoubleType, true),
StructField("petal_width",   DoubleType, true),
StructField("class",  StringType, true)
))

val dataDF = spark.read.format("csv")
             .option("header", "false")
             .schema(irisSchema)
             .load("/files/iris.data")

dataDF.printSchema

root
 |-- sepal_length: double (nullable = true)
 |-- sepal_width: double (nullable = true)
 |-- petal_length: double (nullable = true)
 |-- petal_width: double (nullable = true)
 |-- class: string (nullable = true)

dataDF.show
```

```
+------------+-----------+------------+-----------+-----------+
|sepal_length|sepal_width|petal_length|petal_width|      class|
+------------+-----------+------------+-----------+-----------+
|         5.1|        3.5|         1.4|        0.2|Iris-setosa|
|         4.9|        3.0|         1.4|        0.2|Iris-setosa|
|         4.7|        3.2|         1.3|        0.2|Iris-setosa|
|         4.6|        3.1|         1.5|        0.2|Iris-setosa|
|         5.0|        3.6|         1.4|        0.2|Iris-setosa|
|         5.4|        3.9|         1.7|        0.4|Iris-setosa|
|         4.6|        3.4|         1.4|        0.3|Iris-setosa|
|         5.0|        3.4|         1.5|        0.2|Iris-setosa|
|         4.4|        2.9|         1.4|        0.2|Iris-setosa|
|         4.9|        3.1|         1.5|        0.1|Iris-setosa|
|         5.4|        3.7|         1.5|        0.2|Iris-setosa|
|         4.8|        3.4|         1.6|        0.2|Iris-setosa|
|         4.8|        3.0|         1.4|        0.1|Iris-setosa|
|         4.3|        3.0|         1.1|        0.1|Iris-setosa|
|         5.8|        4.0|         1.2|        0.2|Iris-setosa|
|         5.7|        4.4|         1.5|        0.4|Iris-setosa|
|         5.4|        3.9|         1.3|        0.4|Iris-setosa|
|         5.1|        3.5|         1.4|        0.3|Iris-setosa|
|         5.7|        3.8|         1.7|        0.3|Iris-setosa|
|         5.1|        3.8|         1.5|        0.3|Iris-setosa|
+------------+-----------+------------+-----------+-----------+
only showing top 20 rows

dataDF.describe().show(5,15)

+-------+---------------+---------------+---------------+---------------+
|summary|   sepal_length|    sepal_width|   petal_length|    petal_width|
+-------+---------------+---------------+---------------+---------------+
|  count|            150|            150|            150|            150|
|   mean|5.8433333333...|3.0540000000...|3.7586666666...|1.1986666666...|
| stddev|0.8280661279...|0.4335943113...|1.7644204199...|0.7631607417...|
|    min|            4.3|            2.0|            1.0|            0.1|
|    max|            7.9|            4.4|            6.9|            2.5|
+-------+---------------+---------------+---------------+---------------+
```

```
+--------------+
|         class|
+--------------+
|           150|
|          null|
|          null|
|   Iris-setosa|
|Iris-virginica|
+--------------+

val labelIndexer = new StringIndexer()
                   .setInputCol("class")
                   .setOutputCol("label")

val dataDF2 = labelIndexer.fit(dataDF).transform(dataDF)

dataDF2.printSchema

root
 |-- sepal_length: double (nullable = true)
 |-- sepal_width: double (nullable = true)
 |-- petal_length: double (nullable = true)
 |-- petal_width: double (nullable = true)
 |-- class: string (nullable = true)
 |-- label: double (nullable = false)

dataDF2.show

+------------+-----------+------------+-----------+-----------+-----+
|sepal_length|sepal_width|petal_length|petal_width|      class|label|
+------------+-----------+------------+-----------+-----------+-----+
|         5.1|        3.5|         1.4|        0.2|Iris-setosa|  0.0|
|         4.9|        3.0|         1.4|        0.2|Iris-setosa|  0.0|
|         4.7|        3.2|         1.3|        0.2|Iris-setosa|  0.0|
|         4.6|        3.1|         1.5|        0.2|Iris-setosa|  0.0|
|         5.0|        3.6|         1.4|        0.2|Iris-setosa|  0.0|
|         5.4|        3.9|         1.7|        0.4|Iris-setosa|  0.0|
|         4.6|        3.4|         1.4|        0.3|Iris-setosa|  0.0|
```

```
|         5.0|         3.4|          1.5|         0.2|Iris-setosa|  0.0|
|         4.4|         2.9|          1.4|         0.2|Iris-setosa|  0.0|
|         4.9|         3.1|          1.5|         0.1|Iris-setosa|  0.0|
|         5.4|         3.7|          1.5|         0.2|Iris-setosa|  0.0|
|         4.8|         3.4|          1.6|         0.2|Iris-setosa|  0.0|
|         4.8|         3.0|          1.4|         0.1|Iris-setosa|  0.0|
|         4.3|         3.0|          1.1|         0.1|Iris-setosa|  0.0|
|         5.8|         4.0|          1.2|         0.2|Iris-setosa|  0.0|
|         5.7|         4.4|          1.5|         0.4|Iris-setosa|  0.0|
|         5.4|         3.9|          1.3|         0.4|Iris-setosa|  0.0|
|         5.1|         3.5|          1.4|         0.3|Iris-setosa|  0.0|
|         5.7|         3.8|          1.7|         0.3|Iris-setosa|  0.0|
|         5.1|         3.8|          1.5|         0.3|Iris-setosa|  0.0|
+------------+------------+-------------+------------+-----------+-----+
only showing top 20 rows

import org.apache.spark.ml.feature.VectorAssembler

val features = Array("sepal_length","sepal_width","petal_length",
"petal_width")

val assembler = new VectorAssembler()
                .setInputCols(features)
                .setOutputCol("features")

val dataDF3 = assembler.transform(dataDF2)

dataDF3.printSchema

root
 |-- sepal_length: double (nullable = true)
 |-- sepal_width: double (nullable = true)
 |-- petal_length: double (nullable = true)
 |-- petal_width: double (nullable = true)
 |-- class: string (nullable = true)
 |-- label: double (nullable = false)
 |-- features: vector (nullable = true)
```

```
dataDF3.show

+------------+-----------+------------+-----------+-----------+-----+
|sepal_length|sepal_width|petal_length|petal_width|      class|label|
+------------+-----------+------------+-----------+-----------+-----+
|         5.1|        3.5|         1.4|        0.2|Iris-setosa|  0.0|
|         4.9|        3.0|         1.4|        0.2|Iris-setosa|  0.0|
|         4.7|        3.2|         1.3|        0.2|Iris-setosa|  0.0|
|         4.6|        3.1|         1.5|        0.2|Iris-setosa|  0.0|
|         5.0|        3.6|         1.4|        0.2|Iris-setosa|  0.0|
|         5.4|        3.9|         1.7|        0.4|Iris-setosa|  0.0|
|         4.6|        3.4|         1.4|        0.3|Iris-setosa|  0.0|
|         5.0|        3.4|         1.5|        0.2|Iris-setosa|  0.0|
|         4.4|        2.9|         1.4|        0.2|Iris-setosa|  0.0|
|         4.9|        3.1|         1.5|        0.1|Iris-setosa|  0.0|
|         5.4|        3.7|         1.5|        0.2|Iris-setosa|  0.0|
|         4.8|        3.4|         1.6|        0.2|Iris-setosa|  0.0|
|         4.8|        3.0|         1.4|        0.1|Iris-setosa|  0.0|
|         4.3|        3.0|         1.1|        0.1|Iris-setosa|  0.0|
|         5.8|        4.0|         1.2|        0.2|Iris-setosa|  0.0|
|         5.7|        4.4|         1.5|        0.4|Iris-setosa|  0.0|
|         5.4|        3.9|         1.3|        0.4|Iris-setosa|  0.0|
|         5.1|        3.5|         1.4|        0.3|Iris-setosa|  0.0|
|         5.7|        3.8|         1.7|        0.3|Iris-setosa|  0.0|
|         5.1|        3.8|         1.5|        0.3|Iris-setosa|  0.0|
+------------+-----------+------------+-----------+-----------+-----+
+-----------------+
|         features|
+-----------------+
|[5.1,3.5,1.4,0.2]|
|[4.9,3.0,1.4,0.2]|
|[4.7,3.2,1.3,0.2]|
|[4.6,3.1,1.5,0.2]|
|[5.0,3.6,1.4,0.2]|
|[5.4,3.9,1.7,0.4]|
|[4.6,3.4,1.4,0.3]|
```

```
|[5.0,3.4,1.5,0.2]|
|[4.4,2.9,1.4,0.2]|
|[4.9,3.1,1.5,0.1]|
|[5.4,3.7,1.5,0.2]|
|[4.8,3.4,1.6,0.2]|
|[4.8,3.0,1.4,0.1]|
|[4.3,3.0,1.1,0.1]|
|[5.8,4.0,1.2,0.2]|
|[5.7,4.4,1.5,0.4]|
|[5.4,3.9,1.3,0.4]|
|[5.1,3.5,1.4,0.3]|
|[5.7,3.8,1.7,0.3]|
|[5.1,3.8,1.5,0.3]|
+-----------------+
// We will standardize the four attributes (sepal_length, sepal_width,
// petal_length, and petal_width) using StandardScaler even though they all
// have the same scale and measure the same quantity. As discussed earlier,
// standardization is considered the best practice and is a requirement for
// many algorithms such as PCA to execute optimally.

import org.apache.spark.ml.feature.StandardScaler

val scaler = new StandardScaler()
             .setInputCol("features")
             .setOutputCol("scaledFeatures")
             .setWithStd(true)
             .setWithMean(false)

val dataDF4 = scaler.fit(dataDF3).transform(dataDF3)

dataDF4.printSchema

root
 |-- sepal_length: double (nullable = true)
 |-- sepal_width: double (nullable = true)
 |-- petal_length: double (nullable = true)
 |-- petal_width: double (nullable = true)
 |-- class: string (nullable = true)
```

```
 |-- label: double (nullable = false)
 |-- features: vector (nullable = true)
 |-- scaledFeatures: vector (nullable = true)

// Generate two principal components.

val pca = new PCA()
          .setInputCol("scaledFeatures")
          .setOutputCol("pcaFeatures")
          .setK(2)
          .fit(dataDF4)

val dataDF5 = pca.transform(dataDF4)

dataDF5.printSchema

root
 |-- sepal_length: double (nullable = true)
 |-- sepal_width: double (nullable = true)
 |-- petal_length: double (nullable = true)
 |-- petal_width: double (nullable = true)
 |-- class: string (nullable = true)
 |-- label: double (nullable = false)
 |-- features: vector (nullable = true)
 |-- scaledFeatures: vector (nullable = true)
 |-- pcaFeatures: vector (nullable = true)

dataDF5.select("scaledFeatures","pcaFeatures").show(false)

+--------------------------------------------------------------------------+
|scaledFeatures                                                            |
+--------------------------------------------------------------------------+
|[6.158928408838787,8.072061621390857,0.7934616853039358,0.26206798787142]|
|[5.9174018045706,6.9189099611921625,0.7934616853039358,0.26206798787142] |
|[5.675875200302412,7.38017062527164,0.7367858506393691,0.26206798787142] |
|[5.555111898168318,7.149540293231902,0.8501375199685027,0.26206798787142]|
|[6.038165106704694,8.302691953430596,0.7934616853039358,0.26206798787142]|
|[6.52121831524107,8.99458294954981,0.9634891892976364,0.52413597574284]  |
|[5.555111898168318,7.841431289351117,0.7934616853039358,0.39310198180713]|
```

```
|[6.038165106704694,7.841431289351117,0.8501375199685027,0.26206798787142]|
|[5.313585293900131,6.688279629152423,0.7934616853039358,0.26206798787142]|
|[5.9174018045706,7.149540293231902,0.8501375199685027,0.13103399393571]  |
|[6.52121831524107,8.533322285470334,0.8501375199685027,0.26206798787142] |
|[5.7966385024365055,7.841431289351117,0.9068133546330697,0.262067987871] |
|[5.7966385024365055,6.918909961192l625,0.7934616853039358,0.131033993935]|
|[5.192821991766037,6.9189099611921625,0.6234341813102354,0.1310339939351]|
|[7.004271523777445,9.22521328158955,0.6801100159748021,0.26206798787142] |
|[6.883508221643351,10.147734609748506,0.8501375199685027,0.524135975742] |
|[6.52121831524107,8.99458294954981,0.7367858506393691,0.52413597574284]  |
|[6.158928408838787,8.072061621390857,0.7934616853039358,0.39310198180713]|
|[6.883508221643351,8.763952617510071,0.9634891892976364,0.39310198180713]|
|[6.158928408838787,8.763952617510071,0.8501375199685027,0.39310198180713]|
+-------------------------------------------------------------------------+

+----------------------------------------+
|pcaFeatures                             |
+----------------------------------------+
|[-1.7008636408214346,-9.798112476165109] |
|[-1.8783851549940478,-8.640880678324866] |
|[-1.597800192305247,-8.976683127367169]  |
|[-1.6613406138855684,-8.720650458966217] |
|[-1.5770426874367196,-9.96661148272853]  |
|[-1.8942207975522354,-10.80757533867312] |
|[-1.5202989381570455,-9.368410789070643] |
|[-1.7314610064823877,-9.540884243679617] |
|[-1.6237061774493644,-8.202607301741613] |
|[-1.7764763044699745,-8.846965954487347] |
|[-1.8015813990792064,-10.361118028393015]|
|[-1.6382374187586244,-9.452155017757546] |
|[-1.741187558292187,-8.587346593832775]  |
|[-1.3269417814262463,-8.358947926562632] |
|[-1.7728726239179156,-11.177765120852797]|
|[-1.7138964933624494,-12.00737840334759] |
```

```
|[-1.7624485738747564,-10.80279308233496] |
|[-1.7624485738747564,-10.80279308233496] |
|[-1.7624485738747564,-10.80279308233496] |
|[-1.6257080769316516,-10.44826393443861] |
+-----------------------------------------+
```

As discussed earlier, the Iris dataset has three kinds of flower (Iris Setosa, Iris Versicolor, and Iris Virginica). It has four attributes (sepal length, sepal width, petal length, and petal width). Let's plot the samples on the two principal components. As you can see from Figure 4-8, Iris Setosa is well separated from the other two classes, while Iris Versicolor and Iris Virginica slightly overlap.

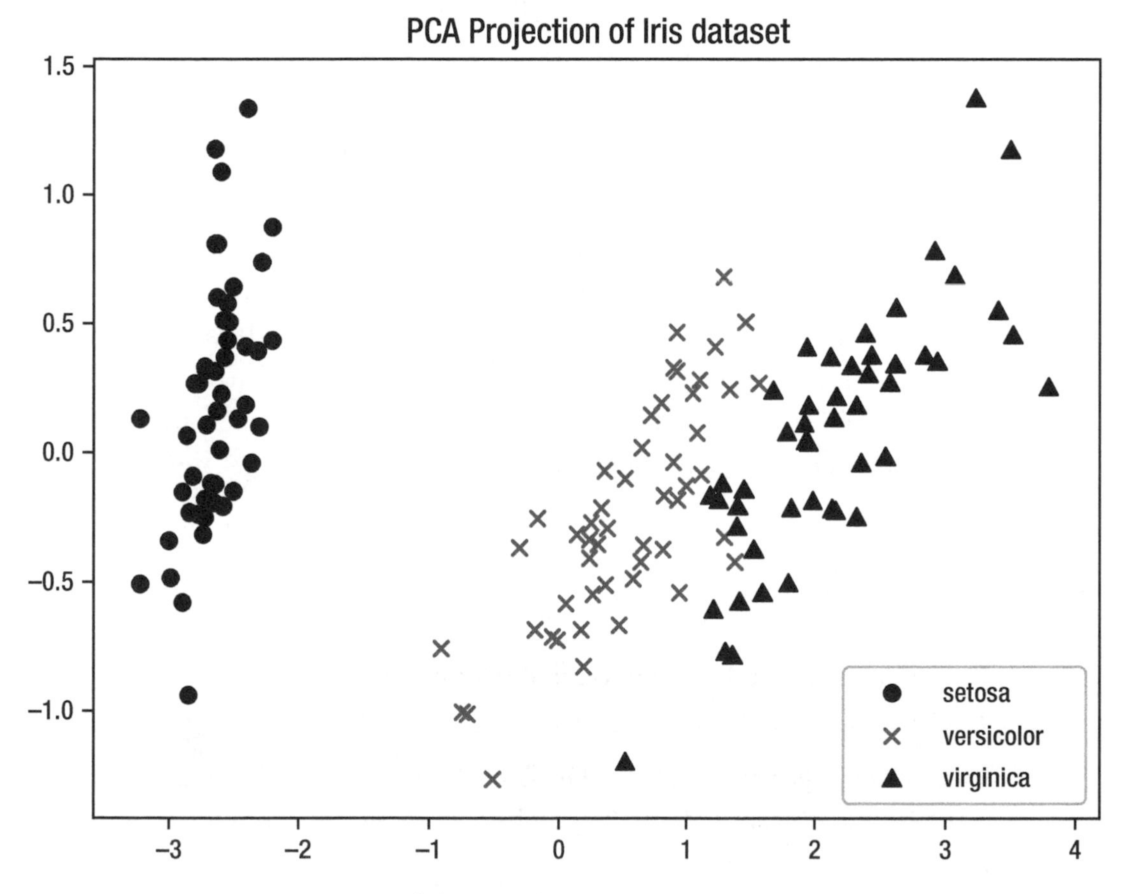

Figure 4-8. *PCA projection of the Iris dataset*

The explainedVariance method returns a vector containing the proportions of variance explained by each principal component. Our goal is to retain as much variance as possible in our new principal components.

```
pca.explainedVariance
res5: org.apache.spark.ml.linalg.DenseVector = [0.7277045209380264,0.230305
23267679512]
```

Based on the output of the method, the first principal component explains 72.77% of the variance, while 23.03% of the variance is explained by the second principal component. Cumulatively, the two principal components explain 95.8% of the variance. As you can see, we lost some information when we reduced our dimensions. This is generally an acceptable trade-off if there is substantial training performance improvement while maintaining good model accuracy.

Summary

We discussed several unsupervised learning techniques and learned how to apply them to real-world business use cases. Unsupervised learning has seen resurgence in popularity in recent years with the advent of big data. Techniques such as clustering, anomaly detection, and principal component analysis help make sense of the deluge of unstructured data generated by mobile and IoT devices, sensors, social media, and more. It is a powerful tool to have in your machine learning arsenal.

References

i. Kurt Vonnegut; "18. The Most Valuable Commodity on Earth," 1998, Cat's Cradle: A Novel

ii. Chris Piech, Andrew Ng, Michael Jordan; "K Means," stanford.edu, 2013, `https://stanford.edu/~cpiech/cs221/handouts/kmeans.html`

iii. Jason Brownlee; "Why One-Hot Encode Data in Machine Learning?", machinelearningmastery.com, 2017, `https://machinelearningmastery.com/why-one-hot-encode-data-in-machine-learning/`

iv. David Talby; "Introducing the Natural Language Processing Library for Apache Spark," Databricks, 2017, https://databricks.com/blog/2017/10/19/introducing-natural-language-processing-library-apache-spark.html

v. Christopher D. Manning et al.; "The Stanford CoreNLP Natural Language Processing Toolkit," Stanford University, https://nlp.stanford.edu/pubs/StanfordCoreNlp2014.pdf

vi. John Snow Labs; "Quick Start," John Snow Labs, 2019, https://nlp.johnsnowlabs.com/docs/en/quickstart

vii. Shivam Bansal; "Ultimate Guide to Understand & Implement Natural Language Processing," Analytics Vidhya, 2017, www.analyticsvidhya.com/blog/2017/01/ultimate-guide-to-understand-implement-natural-language-processing-codes-in-python/

viii. Sebastian Raschka; "Naïve Bayes and Text Classification – Introduction and Theory," sebastiantraschka.com, 2014, https://sebastianraschka.com/Articles/2014_naive_bayes_1.html

ix. Zygmunt Zawadzki; "Get topics words from the LDA model," zstat.pl, 2018, www.zstat.pl/2018/02/07/scala-spark-get-topics-words-from-lda-model/

x. Fei Tony Liu, Kai Ming Ting, Zhia-Hua Zhou; "Isolation Forest," acm.org, 2008, https://dl.acm.org/citation.cfm?id=1511387

xi. Alejandro Correa Bahnsen; "Benefits of Anomaly Detection Using Isolation Forests," easysol.net, 2016, https://blog.easysol.net/using-isolation-forests-anamoly-detection/

xii. Li Sun, et. al.; "Detecting Anomalous User Behavior Using an Extended Isolation Forest Algorithm: An Enterprise Case Study," arxiv.org, 2016, https://arxiv.org/pdf/1609.06676.pdf

xiii. Li Sun, et. al.; "Detecting Anomalous User Behavior Using an Extended Isolation Forest Algorithm: An Enterprise Case Study," arxiv.org, 2016, https://arxiv.org/pdf/1609.06676.pdf

xiv. Zhi-Min Zhang; "Representative subset selection and outlier detection via Isolation Forest," github.com, 2016, https://github.com/zmzhang/IOS

xv. Alejandro Correa Bahnsen; "Benefits of Anomaly Detection Using Isolation Forests," easysol.net, 2016, https://blog.easysol.net/using-isolation-forests-anamoly-detection/

xvi. Fangzhou Yang and contributors; "spark-iforest," github.com, 2018, https://github.com/titicaca/spark-iforest

xvii. Fangzhou Yang and contributors; "spark-iforest," github.com, 2018, https://github.com/titicaca/spark-iforest

xviii. Dr. William H. Wolberg, et al.; "Breast Cancer Wisconsin (Diagnostic) Data Set," archive.isc.uci.edu, 1995, http://archive.ics.uci.edu/ml/datasets/breast+cancer+wisconsin+(diagnostic)

CHAPTER 5

Recommendations

> *Man is the creature who does not know what to desire, and he turns to others in order to make up his mind. We desire what others desire because we imitate their desires.*
>
> —René Girard[i]

Providing personalized recommendations is one of the most popular applications of machine learning. Almost every major retailer, such as Amazon, Alibaba, Walmart, and Target, provides some sort of personalized recommendation based on customer behavior. Streaming services such as Netflix, Hulu, and Spotify provide movie or music recommendations based on user tastes and preferences.

Recommendations are vital in improving customer satisfaction and engagement, which ultimately increases sales and revenue. To highlight the importance of recommendations, 44% of Amazon customers buy from product recommendations they see on Amazon.[ii] A McKinsey report found that 35% of customer sales come directly from Amazon's recommendations. The same study reports that 75% of what viewers watch on Netflix come from personalized recommendations.[iii] Netflix's Chief Product Officer proclaimed in an interview that Netflix's personalized movie and TV show recommendations are worth $1 billion per year to the company.[iv] Alibaba's recommendation engine helped drive record sales, becoming one of the world's largest e-commerce companies and retailing juggernaut with $248 billion in sales in 2013 (more than Amazon and eBay combined).[v]

Recommendations are not limited to retailers and streaming services. Banks use recommendation engines as a targeted marketing tool, using it to offer online banking customers financial products and services such as home or student loans based on their demographic and psychographic profile. Advertising and marketing agencies use recommendation engines to display highly targeted online ads.

B. Quinto, *Next-Generation Machine Learning with Spark*, https://doi.org/10.1007/978-1-4842-5669-5_5

Types of Recommendation Engines

There are several types of recommendation engines.[vi] We will discuss the most popular: collaborative filtering, content-based filtering, and association rules.

Collaborative Filtering with Alternating Least Squares

Collaborative filtering is frequently used to provide personalized recommendation on the Web. Companies that utilize collaborative filtering include Netflix, Amazon, Alibaba, Spotify, and Apple, to mention a few. Collaborative filtering provides recommendations (filtering) based on other people's (collaborative) preferences or tastes. It is based on the idea that people with the same preference are more likely to have the same interests in the future. For instance, Laura likes *Titanic, Apollo 13,* and *The Towering Inferno*. Tom likes *Apollo 13* and *The Towering Inferno*. If Anne likes *Apollo 13,* and based on our calculation anyone who likes *Apollo 13* must also like *The Towering Inferno*, then *The Towering Inferno* could be a potential recommendation to Anne. A product could be any item such as a movie, a song, a video, or a book.

	Movie 1	Movie 2	Movie 3
Laura		4	4
Anne	2	5	?
Tom		4	5

Figure 5-1. *ALS rating matrix*[vii]

Spark MLlib includes a popular algorithm for collaborative filtering called alternating least squares (ALS). ALS models the rating matrix (Figure 5-1) as the product of users and product factors[viii] (Figure 5-2). ALS utilizes the least squares computation to minimize the estimation errors,[ix] iterating alternatively between fixing customers factors and solving for product factors and vice versa until the process converges. Spark MLlib implements a blocked version of ALS that leverages Spark's distributed processing capabilities by grouping the two sets of factors (referred to as "users" and "products") into blocks and reduces communication by only sending one copy of each user vector to each product block on each iteration, and only for the product blocks that need that user's feature vector.[x]

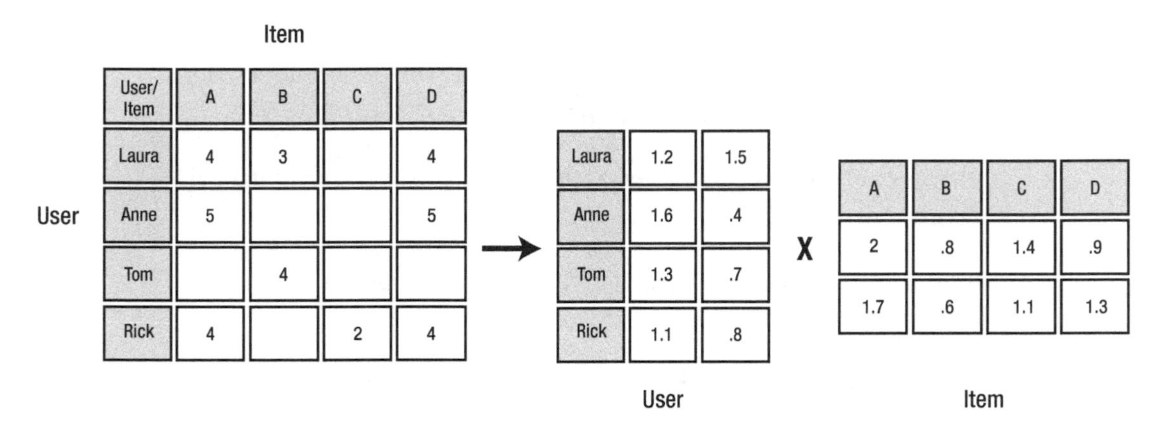

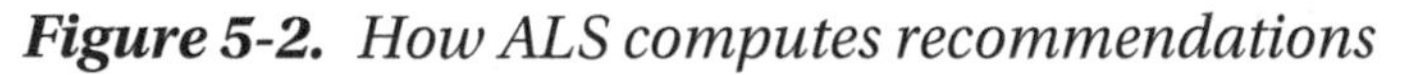

Figure 5-2. *How ALS computes recommendations*

Spark MLlib's ALS implementation supports both explicit and implicit ratings. Explicit ratings (default) require that a user's rating for a product to be a score (e.g., 1–5 thumbs up), while implicit ratings signify confidence that users will interact with a product (e.g., number of clicks or page views, or the number of times a video has been streamed). Implicit ratings are more common in real-life scenarios since not every company gathers explicit ratings for their products. Implicit ratings however can be extracted from corporate data such as web logs, viewing habits, or sales transactions. Spark MLlib's ALS implementation uses integers for item and user ids, which means item and user ids must be within the range of value of integers and a maximum value of 2,147,483,647.

Note Alternating least squares (ALS) is described in the paper "Scalable Collaborative Filtering with Jointly Derived Neighborhood Interpolation Weights" by Yehuda Koren and Robert M. Bell.[xi]

Parameters

Spark MLlib's ALS implementation supports the following parameters[xii]:

- *alpha:* Applicable to the implicit feedback version of ALS that directs the baseline confidence in preference observations
- *numBlocks: U*sed for parallel processing; the number of blocks the items and users will be partitioned into
- *nonnegative:* Indicates whether or not to use nonnegative constraints for least squares
- *implicitPrefs:* Indicates whether to use explicit feedback or implicit feedback
- *k:* Indicates the number of latent factors in the model
- *regParam*: The regularization parameter
- *maxIter:* Indicates the maximum number of iterations to execute.

Example

We will use the MovieLens dataset to build a toy movie recommendation system. The dataset can be downloaded from `https://grouplens.org/datasets/movielens/`. There are multiple files included in the dataset, but we're mainly interested in ratings.csv. Each row in the file shown in Listing 5-1 contains a user's explicit rating for a movie (1-5 rating).

Listing 5-1. Movie Recommendations with ALS

```
val dataDF = spark.read.option("header", "true")
             .option("inferSchema", "true")
             .csv("ratings.csv")

dataDF.printSchema
root
 |-- userId: integer (nullable = true)
 |-- movieId: integer (nullable = true)
 |-- rating: double (nullable = true)
 |-- timestamp: integer (nullable = true)
```

```
dataDF.show
+------+-------+------+---------+
|userId|movieId|rating|timestamp|
+------+-------+------+---------+
|     1|      1|   4.0|964982703|
|     1|      3|   4.0|964981247|
|     1|      6|   4.0|964982224|
|     1|     47|   5.0|964983815|
|     1|     50|   5.0|964982931|
|     1|     70|   3.0|964982400|
|     1|    101|   5.0|964980868|
|     1|    110|   4.0|964982176|
|     1|    151|   5.0|964984041|
|     1|    157|   5.0|964984100|
|     1|    163|   5.0|964983650|
|     1|    216|   5.0|964981208|
|     1|    223|   3.0|964980985|
|     1|    231|   5.0|964981179|
|     1|    235|   4.0|964980908|
|     1|    260|   5.0|964981680|
|     1|    296|   3.0|964982967|
|     1|    316|   3.0|964982310|
|     1|    333|   5.0|964981179|
|     1|    349|   4.0|964982563|
+------+-------+------+---------+
only showing top 20 rows

val Array(trainingData, testData) = dataDF.randomSplit(Array(0.7, 0.3))

import org.apache.spark.ml.recommendation.ALS

val als = new ALS()
          .setMaxIter(15)
          .setRank(10)
          .setSeed(1234)
          .setRatingCol("rating")
          .setUserCol("userId")
          .setItemCol("movieId")
```

```
val model = als.fit(trainingData)

val predictions = model.transform(testData)

predictions.printSchema
root
 |-- userId: integer (nullable = true)
 |-- movieId: integer (nullable = true)
 |-- rating: double (nullable = true)
 |-- timestamp: integer (nullable = true)
 |-- prediction: float (nullable = false)

predictions.show

+------+-------+------+----------+----------+
|userId|movieId|rating| timestamp|prediction|
+------+-------+------+----------+----------+
|   133|    471|   4.0| 843491793| 2.5253267|
|   602|    471|   4.0| 840876085| 3.2802277|
|   182|    471|   4.5|1054779644| 3.6534667|
|   500|    471|   1.0|1005528017| 3.5033386|
|   387|    471|   3.0|1139047519| 2.6689813|
|   610|    471|   4.0|1479544381|  3.006948|
|   136|    471|   4.0| 832450058| 3.1404104|
|   312|    471|   4.0|1043175564|  3.109232|
|   287|    471|   4.5|1110231536| 2.9776838|
|    32|    471|   3.0| 856737165| 3.5183017|
|   469|    471|   5.0| 965425364| 2.8298397|
|   608|    471|   1.5|1117161794|  3.007364|
|   373|    471|   5.0| 846830388| 3.9275675|
|   191|    496|   5.0| 829760898|       NaN|
|    44|    833|   2.0| 869252237| 2.4776468|
|   609|    833|   3.0| 847221080| 1.9167987|
|   608|    833|   0.5|1117506344|  2.220617|
|   463|   1088|   3.5|1145460096| 3.0794377|
|    47|   1088|   4.0|1496205519| 2.4831696|
|   479|   1088|   4.0|1039362157| 3.5400867|
+------+-------+------+----------+----------+
```

```
import org.apache.spark.ml.evaluation.RegressionEvaluator

val evaluator = new RegressionEvaluator()
                .setPredictionCol("prediction")
                .setLabelCol("rating")
                .setMetricName("rmse")

val rmse = evaluator.evaluate(predictions)
rmse: Double = NaN
```

It looks like the evaluator doesn't like NaN values in the prediction DataFrame. Let's fix it for now by removing rows with NaN values. We discuss how to use the coldStartStrategy parameter to handle this issue later.

```
val predictions2 = predictions.na.drop

predictions2.show

+------+-------+------+----------+----------+
|userId|movieId|rating| timestamp|prediction|
+------+-------+------+----------+----------+
|   133|    471|   4.0| 843491793| 2.5253267|
|   602|    471|   4.0| 840876085| 3.2802277|
|   182|    471|   4.5|1054779644| 3.6534667|
|   500|    471|   1.0|1005528017| 3.5033386|
|   387|    471|   3.0|1139047519| 2.6689813|
|   610|    471|   4.0|1479544381|  3.006948|
|   136|    471|   4.0| 832450058| 3.1404104|
|   312|    471|   4.0|1043175564|  3.109232|
|   287|    471|   4.5|1110231536| 2.9776838|
|    32|    471|   3.0| 856737165| 3.5183017|
|   469|    471|   5.0| 965425364| 2.8298397|
|   608|    471|   1.5|1117161794|  3.007364|
|   373|    471|   5.0| 846830388| 3.9275675|
|    44|    833|   2.0| 869252237| 2.4776468|
|   609|    833|   3.0| 847221080| 1.9167987|
|   608|    833|   0.5|1117506344|  2.220617|
|   463|   1088|   3.5|1145460096| 3.0794377|
|    47|   1088|   4.0|1496205519| 2.4831696|
```

```
|   479|   1088|   4.0|1039362157| 3.5400867|
|   554|   1088|   5.0| 944900489| 3.3577442|
+------+-------+-----+---------+----------+
only showing top 20 rows

val evaluator = new RegressionEvaluator()
                .setPredictionCol("prediction")
                .setLabelCol("rating")
                .setMetricName("rmse")

val rmse = evaluator.evaluate(predictions2)
rmse: Double = 0.9006479893684061
```

Note When using ALS, you will sometimes encounter users and/or items in your test dataset that were not present when the model was trained. New users or items may not have any ratings and on which the model has not been trained. This is known as the cold start problem. You may also encounter this when the data is randomly split between evaluation and training datasets. Predictions are set to NaN when a user and/or item is not in the model. This is the reason why we encountered a NaN result earlier when evaluating the model. To address this concern, Spark provides a coldStartStrategy parameter that can be set to drop all rows in the prediction DataFrame that contain NaN values.[xiii]

Let's generate some recommendations.

Recommend top three movies for all users.

```
model.recommendForAllUsers(3).show(false)

+------+---------------------------------------------------------+
|userId|recommendations                                          |
+------+---------------------------------------------------------+
|471   |[[7008, 4.8596725], [7767, 4.8047066], [26810, 4.7513227]]|
|463   |[[33649, 5.0881286], [3347, 4.7693057], [68945, 4.691733]]|
|496   |[[6380, 4.946864], [26171, 4.8910613], [7767, 4.868356]]  |
|148   |[[183897, 4.972257], [6732, 4.561547], [33649, 4.5440807]]|
|540   |[[26133, 5.19643], [68945, 5.1259947], [3379, 5.1259947]] |
|392   |[[3030, 6.040107], [4794, 5.6566052], [55363, 5.4429026]] |
```

```
|243   |[[1223, 6.5019746], [68945, 6.353135], [3379, 6.353135]]  |
|31    |[[4256, 5.3734074], [49347, 5.365612], [7071, 5.3175936]] |
|516   |[[4429, 4.8486495], [48322, 4.8443394], [28, 4.8082485]]  |
|580   |[[86347, 5.20571], [4256, 5.0522637], [72171, 5.037114]]  |
|251   |[[33649, 5.6993585], [68945, 5.613014], [3379, 5.613014]] |
|451   |[[68945, 5.392536], [3379, 5.392536], [905, 5.336588]]    |
|85    |[[25771, 5.2532864], [8477, 5.186757], [99764, 5.1611686]]|
|137   |[[7008, 4.8952146], [26131, 4.8543305], [3200, 4.6918836]]|
|65    |[[33649, 4.695069], [3347, 4.5379376], [7071, 4.535537]]  |
|458   |[[3404, 5.7415047], [7018, 5.390625], [42730, 5.343014]]  |
|481   |[[232, 4.393473], [3473, 4.3804317], [26133, 4.357505]]   |
|53    |[[3200, 6.5110188], [33649, 6.4942613], [3347, 6.452143]] |
|255   |[[86377, 5.9217377], [5047, 5.184309], [6625, 4.962062]]  |
|588   |[[26133, 4.7600465], [6666, 4.65716], [39444, 4.613207]]  |
+------+----------------------------------------------------------+
only showing top 20 rows
```

Recommend top three users for all movies.

```
model.recommendForAllItems(3).show(false)

+-------+----------------------------------------------------+
|movieId|recommendations                                     |
+-------+----------------------------------------------------+
|1580   |[[53, 4.939177], [543, 4.8362885], [452, 4.5791063]] |
|4900   |[[147, 3.0081954], [375, 2.9420073], [377, 2.6285374]]|
|5300   |[[53, 4.29147], [171, 4.129584], [375, 4.1011653]]    |
|6620   |[[392, 5.0614614], [191, 4.820595], [547, 4.7811346]] |
|7340   |[[413, 3.2256641], [578, 3.1126869], [90, 3.0790782]] |
|32460  |[[53, 5.642673], [12, 5.5260286], [371, 5.2030106]]   |
|54190  |[[53, 5.544555], [243, 5.486003], [544, 5.243029]]    |
|471    |[[51, 5.073474], [53, 4.8641024], [337, 4.656805]]    |
|1591   |[[112, 4.250576], [335, 4.147236], [207, 4.05843]]    |
|140541 |[[393, 4.4335465], [536, 4.1968756], [388, 4.0388694]]|
|1342   |[[375, 4.3189483], [313, 3.663758], [53, 3.5866988]]  |
|2122   |[[375, 4.3286233], [147, 4.3245177], [112, 3.8350344]]|
|2142   |[[51, 3.9718416], [375, 3.8228302], [122, 3.8117828]] |
```

```
|7982   |[[191, 5.297085], [547, 5.020829], [187, 4.984965]]   |
|44022  |[[12, 4.5919843], [53, 4.501897], [523, 4.301981]]    |
|141422 |[[456, 2.7050805], [597, 2.6988854], [498, 2.6347125]]|
|833    |[[53, 3.8047972], [543, 3.740805], [12, 3.6920836]]   |
|5803   |[[537, 3.8269677], [544, 3.8034997], [259, 3.76062]]  |
|7993   |[[375, 2.93635], [53, 2.9159238], [191, 2.8663528]]   |
|160563 |[[53, 4.048704], [243, 3.9232922], [337, 3.7616432]]  |
+-------+------------------------------------------------------+
only showing top 20 rows
```

Generate top three user recommendations for a specified set of movies.

```
model.recommendForItemSubset(Seq((111), (202), (225), (347), (488)).
toDF("movieId"), 3).show(false)

+-------+----------------------------------------------------+
|movieId|recommendations                                     |
+-------+----------------------------------------------------+
|225    |[[53, 4.4893017], [147, 4.483344], [276, 4.2529426]]|
|111    |[[375, 5.113064], [53, 4.9947076], [236, 4.9493203]]|
|347    |[[191, 4.686208], [236, 4.51165], [40, 4.409832]]   |
|202    |[[53, 3.349618], [578, 3.255436], [224, 3.245058]]  |
|488    |[[558, 3.3870435], [99, 3.2978806], [12, 3.2749753]]|
+-------+----------------------------------------------------+
```

Generate top three movie recommendations for a specified set of users.

```
model.recommendForUserSubset(Seq((111), (100), (110), (120), (130)).
toDF("userId"), 3).show(false)

+------+--------------------------------------------------------------+
|userId|recommendations                                               |
+------+--------------------------------------------------------------+
|111   |[[106100, 4.956068], [128914, 4.9050474], [162344, 4.9050474]]|
|120   |[[26865, 4.979374], [3508, 4.6825113], [3200, 4.6406555]]     |
|100   |[[42730, 5.2531567], [5867, 5.1075697], [3404, 5.0877166]]    |
|130   |[[86377, 5.224841], [3525, 5.0586476], [92535, 4.9758487]]    |
|110   |[[49932, 4.6330786], [7767, 4.600622], [26171, 4.5615706]]    |
+------+--------------------------------------------------------------+
```

Collaborative filtering can be very effective in providing highly relevant recommendations. It scales well and can handle extremely large datasets. For collaborative filtering to operate optimally, it needs access to a large of amount of data. The more data, the better. As time progresses and ratings start to accumulate, recommendations become more and more accurate. Access to large datasets is often a problem during the early stages of implementation. One solution is to use content-based filtering in conjunction with collaborative filtering. Since content-based filtering doesn't rely on user activity, it can immediately start providing recommendations, gradually increasing your dataset over time.

Market Basket Analysis with FP-Growth

Market basket analysis is a simple but important technique commonly used by retailers to provide product recommendations. It uses transactional datasets to determine which products are frequently purchased together. Retailers can use the recommendations to inform personalized cross-selling and upselling, helping increase conversion and maximizing the value of each customer.

You most likely have already seen market basket analysis in action while browsing through Amazon.com. An Amazon.com product page will usually have a section called "Customers who bought this item also bought," presenting you with a list of items that are frequently bought together with the product you are currently browsing. That list is generated via market basket analysis. Market basket analysis is also used by brick-and-mortar retailers for store optimization by informing product placements and adjacencies in planograms. The idea is to drive more sales by placing complementary items next to each other.

Market basket analysis uses association rules learning to make recommendations. Association rules look for relationships between items using large transactional datasets.[xiv] Association rules are calculated from two or more items called *itemsets*. An association rule consists of an antecedent (if) and a consequent (then). For example, if someone buys cookies (antecedent), then the person is also more likely to buy milk (consequent). Popular association rule algorithms include Apriori, SETM, ECLAT, and FP-Growth. Spark MLlib includes a highly scalable implementation of FP-Growth for association rule mining.[xv] FP-Growth identifies frequent items and calculates item frequencies using a frequent pattern (*"FP"* stands for frequent pattern) tree structure.[xvi]

Note FP-Growth is described in the paper "*Mining Frequent Patterns Without Candidate Generation*" by Jiawei Han, Jian Pei, and Iwen Yin.[xvii]

Example

We will use the popular Instacart Online Grocery Shopping Dataset for our market basket analysis example using FP-Growth.[xviii] The dataset contains 3.4 million grocery orders for 50,000 products from 200,000 Instacart customers. You can download the dataset from `www.instacart.com/datasets/grocery-shopping-2017`. For FP-Growth, we only need the products and order_products_train tables (see Listing 5-2).

Listing 5-2. Market Basket Analysis with FP-Growth

```
val productsDF = spark.read.format("csv")
                 .option("header", "true")
                 .option("inferSchema","true")
                 .load("/instacart/products.csv")

ProductsDF.show(false)

+----------+--------------------------------------------------+
|product_id|product_name                                      |
+----------+--------------------------------------------------+
|1         |Chocolate Sandwich Cookies                        |
|2         |All-Seasons Salt                                  |
|3         |Robust Golden Unsweetened Oolong Tea              |
|4         |Smart Ones Classic Favorites Mini Rigatoni With   |
|5         |Green Chile Anytime Sauce                         |
|6         |Dry Nose Oil                                      |
|7         |Pure Coconut Water With Orange                    |
|8         |Cut Russet Potatoes Steam N' Mash                 |
|9         |Light Strawberry Blueberry Yogurt                 |
|10        |Sparkling Orange Juice & Prickly Pear Beverage    |
|11        |Peach Mango Juice                                 |
|12        |Chocolate Fudge Layer Cake                        |
```

```
|13        |Saline Nasal Mist                                |
|14        |Fresh Scent Dishwasher Cleaner                   |
|15        |Overnight Diapers Size 6                         |
|16        |Mint Chocolate Flavored Syrup                    |
|17        |Rendered Duck Fat                                |
|18        |Pizza for One Suprema  Frozen Pizza              |
|19        |Gluten Free Quinoa Three Cheese & Mushroom Blend |
|20        |Pomegranate Cranberry & Aloe Vera Enrich Drink   |
+----------+-------------------------------------------------+

+--------+-------------+
|aisle_id|department_id|
+--------+-------------+
|61      |19           |
|104     |13           |
|94      |7            |
|38      |1            |
|5       |13           |
|11      |11           |
|98      |7            |
|116     |1            |
|120     |16           |
|115     |7            |
|31      |7            |
|119     |1            |
|11      |11           |
|74      |17           |
|56      |18           |
|103     |19           |
|35      |12           |
|79      |1            |
|63      |9            |
|98      |7            |
+--------+-------------+

only showing top 20 rows
```

```
val orderProductsDF = spark.read.format("csv")
                      .option("header", "true")
                      .option("inferSchema","true")
                      .load("/instacart/order_products__train.csv")

orderProductsDF.show()
+--------+----------+-----------------+---------+
|order_id|product_id|add_to_cart_order|reordered|
+--------+----------+-----------------+---------+
|       1|     49302|                1|        1|
|       1|     11109|                2|        1|
|       1|     10246|                3|        0|
|       1|     49683|                4|        0|
|       1|     43633|                5|        1|
|       1|     13176|                6|        0|
|       1|     47209|                7|        0|
|       1|     22035|                8|        1|
|      36|     39612|                1|        0|
|      36|     19660|                2|        1|
|      36|     49235|                3|        0|
|      36|     43086|                4|        1|
|      36|     46620|                5|        1|
|      36|     34497|                6|        1|
|      36|     48679|                7|        1|
|      36|     46979|                8|        1|
|      38|     11913|                1|        0|
|      38|     18159|                2|        0|
|      38|      4461|                3|        0|
|      38|     21616|                4|        1|
+--------+----------+-----------------+---------+
only showing top 20 rows

// Create temporary tables.

orderProductsDF.createOrReplaceTempView("order_products_train")
productsDF.createOrReplaceTempView("products")
```

```
val joinedData = spark.sql("select p.product_name, o.order_id from order_
products_train o inner join products p where p.product_id = o.product_id")

import org.apache.spark.sql.functions.max
import org.apache.spark.sql.functions.collect_set

val basketsDF = joinedData
                .groupBy("order_id")
                .agg(collect_set("product_name")
                .alias("items"))

basketsDF.createOrReplaceTempView("baskets"

basketsDF.show(20,55)

+--------+-------------------------------------------------------+
|order_id|                                                  items|
+--------+-------------------------------------------------------+
|    1342|[Raw Shrimp, Seedless Cucumbers, Versatile Stain Rem...|
|    1591|[Cracked Wheat, Strawberry Rhubarb Yoghurt, Organic ...|
|    4519|[Beet Apple Carrot Lemon Ginger Organic Cold Pressed...|
|    4935|                                                [Vodka]|
|    6357|[Globe Eggplant, Panko Bread Crumbs, Fresh Mozzarell...|
|   10362|[Organic Baby Spinach, Organic Spring Mix, Organic L...|
|   19204|[Reduced Fat Crackers, Dishwasher Cleaner, Peanut Po...|
|   29601|[Organic Red Onion, Small Batch Authentic Taqueria T...|
|   31035|[Organic Cripps Pink Apples, Organic Golden Deliciou...|
|   40011|[Organic Baby Spinach, Organic Blues Bread with Blue...|
|   46266|[Uncured Beef Hot Dog, Organic Baby Spinach, Smoked ...|
|   51607|[Donut House Chocolate Glazed Donut Coffee K Cup, Ma...|
|   58797|[Concentrated Butcher's Bone Broth, Chicken, Seedles...|
|   61793|[Raspberries, Green Seedless Grapes, Clementines, Na...|
|   67089|[Original Tofurky Deli Slices, Sharp Cheddar Cheese,...|
|   70863|[Extra Hold Non-Aerosol Hair Spray, Bathroom Tissue,...|
|   88674|[Organic Coconut Milk, Everything Bagels, Rosemary, ...|
|   91937|[No. 485 Gin, Monterey Jack Sliced Cheese, Tradition...|
|   92317|[Red Vine Tomato, Harvest Casserole Bowls, Organic B...|
|   99621|[Organic Baby Arugula, Organic Garlic, Fennel, Lemon...|
+--------+-------------------------------------------------------+
```

```
only showing top 20 rows

import org.apache.spark.ml.fpm.FPGrowth

// FPGrowth only needs a string containing the list of items.

val basketsDF = spark.sql("select items from baskets")
               .as[Array[String]].toDF("items")

basketsDF.show(20,55)
+-------------------------------------------------------+
|                                                  items|
+-------------------------------------------------------+
|[Raw Shrimp, Seedless Cucumbers, Versatile Stain Rem...|
|[Cracked Wheat, Strawberry Rhubarb Yoghurt, Organic ...|
|[Beet Apple Carrot Lemon Ginger Organic Cold Pressed...|
|                                                [Vodka]|
|[Globe Eggplant, Panko Bread Crumbs, Fresh Mozzarell...|
|[Organic Baby Spinach, Organic Spring Mix, Organic L...|
|[Reduced Fat Crackers, Dishwasher Cleaner, Peanut Po...|
|[Organic Red Onion, Small Batch Authentic Taqueria T...|
|[Organic Cripps Pink Apples, Organic Golden Deliciou...|
|[Organic Baby Spinach, Organic Blues Bread with Blue...|
|[Uncured Beef Hot Dog, Organic Baby Spinach, Smoked ...|
|[Donut House Chocolate Glazed Donut Coffee K Cup, Ma...|
|[Concentrated Butcher's Bone Broth, Chicken, Seedles...|
|[Raspberries, Green Seedless Grapes, Clementines, Na...|
|[Original Tofurky Deli Slices, Sharp Cheddar Cheese,...|
|[Extra Hold Non-Aerosol Hair Spray, Bathroom Tissue,...|
|[Organic Coconut Milk, Everything Bagels, Rosemary, ...|
|[No. 485 Gin, Monterey Jack Sliced Cheese, Tradition...|
|[Red Vine Tomato, Harvest Casserole Bowls, Organic B...|
|[Organic Baby Arugula, Organic Garlic, Fennel, Lemon...|
+-------------------------------------------------------+
only showing top 20 rows
```

```
val fpgrowth = new FPGrowth()
              .setItemsCol("items")
              .setMinSupport(0.002)
              .setMinConfidence(0)

val model = fpgrowth.fit(basketsDF)

// Frequent itemsets.

val mostPopularItems = model.freqItemsets

mostPopularItems.createOrReplaceTempView("mostPopularItems")

// Verify results.

spark.sql("select * from mostPopularItems wheresize(items) >= 2 order by
freq desc")
          .show(20,55)
+-----------------------------------------------+----+
|                                          items|freq|
+-----------------------------------------------+----+
|[Organic Strawberries, Bag of Organic Bananas]|3074|
|[Organic Hass Avocado, Bag of Organic Bananas]|2420|
|[Organic Baby Spinach, Bag of Organic Bananas]|2236|
|                     [Organic Avocado, Banana]|2216|
|                [Organic Strawberries, Banana]|2174|
|                         [Large Lemon, Banana]|2158|
|                [Organic Baby Spinach, Banana]|2000|
|                        [Strawberries, Banana]|1948|
| [Organic Raspberries, Bag of Organic Bananas]|1780|
|   [Organic Raspberries, Organic Strawberries]|1670|
|  [Organic Baby Spinach, Organic Strawberries]|1639|
|                          [Limes, Large Lemon]|1595|
|  [Organic Hass Avocado, Organic Strawberries]|1539|
|      [Organic Avocado, Organic Baby Spinach]|1402|
|                [Organic Avocado, Large Lemon]|1349|
|                               [Limes, Banana]|1331|
```

```
|   [Organic Blueberries, Organic Strawberries]|1269|
|    [Organic Cucumber, Bag of Organic Bananas]|1268|
|  [Organic Hass Avocado, Organic Baby Spinach]|1252|
|           [Large Lemon, Organic Baby Spinach]|1238|
+----------------------------------------------+----+
only showing top 20 rows

spark.sql("select * from mostPopularItems where
          size(items) > 2 order by freq desc")
          .show(20,65)

+-----------------------------------------------------------------+----+
|                                                            items|freq|
+-----------------------------------------------------------------+----+
|[Organic Hass Avocado, Organic Strawberries, Bag of Organic Ba...| 710|
|[Organic Raspberries, Organic Strawberries, Bag of Organic Ban...| 649|
|[Organic Baby Spinach, Organic Strawberries, Bag of Organic Ba...| 587|
|[Organic Raspberries, Organic Hass Avocado, Bag of Organic Ban...| 531|
|[Organic Hass Avocado, Organic Baby Spinach, Bag of Organic Ba...| 497|
|                  [Organic Avocado, Organic Baby Spinach, Banana]| 484|
|                           [Organic Avocado, Large Lemon, Banana]| 477|
|                                     [Limes, Large Lemon, Banana]| 452|
| [Organic Cucumber, Organic Strawberries, Bag of Organic Bananas]| 424|
|                            [Limes, Organic Avocado, Large Lemon]| 389|
|[Organic Raspberries, Organic Hass Avocado, Organic Strawberries]| 381|
|                  [Organic Avocado, Organic Strawberries, Banana]| 379|
|             [Organic Baby Spinach, Organic Strawberries, Banana]| 376|
|[Organic Blueberries, Organic Strawberries, Bag of Organic Ban...| 374|
|                      [Large Lemon, Organic Baby Spinach, Banana]| 371|
| [Organic Cucumber, Organic Hass Avocado, Bag of Organic Bananas]| 366|
|    [Organic Lemon, Organic Hass Avocado, Bag of Organic Bananas]| 353|
|                                 [Limes, Organic Avocado, Banana]| 352|
|[Organic Whole Milk, Organic Strawberries, Bag of Organic Bana...| 339|
|             [Organic Avocado, Large Lemon, Organic Baby Spinach]| 334|
+-----------------------------------------------------------------+----+
only showing top 20 rows
```

Shown are items that are most likely to be purchased together. The most popular in the list is a combination of organic avocado, organic strawberries, and a bag of organic bananas. This kind of lists could be the basis for "frequently bought together"-type recommendations.

```
// The FP-Growth model also generates association rules. The output includes
// the antecedent, consequent, and confidence (probability). The minimum
// confidence for generating association rule is determined by the
// minConfidence parameter.

val AssocRules = model.associationRules
AssocRules.createOrReplaceTempView("AssocRules")

spark.sql("select antecedent, consequent,
          confidence from AssocRules order by confidence desc")
          .show(20,55)

+-------------------------------------------------------+
|                                             antecedent|
+-------------------------------------------------------+
|            [Organic Raspberries, Organic Hass Avocado]|
|                        [Strawberries, Organic Avocado]|
|           [Organic Hass Avocado, Organic Strawberries]|
|                 [Organic Lemon, Organic Hass Avocado]|
|                 [Organic Lemon, Organic Strawberries]|
|              [Organic Cucumber, Organic Hass Avocado]|
|[Organic Large Extra Fancy Fuji Apple, Organic Straw...|
|           [Organic Yellow Onion, Organic Hass Avocado]|
|                            [Strawberries, Large Lemon]|
|            [Organic Blueberries, Organic Raspberries]|
|              [Organic Cucumber, Organic Strawberries]|
|              [Organic Zucchini, Organic Hass Avocado]|
|            [Organic Raspberries, Organic Baby Spinach]|
|           [Organic Hass Avocado, Organic Baby Spinach]|
|              [Organic Zucchini, Organic Strawberries]|
|            [Organic Raspberries, Organic Strawberries]|
```

```
|                                      [Bartlett Pears]|
|                                         [Gala Apples]|
|                           [Limes, Organic Avocado]|
|            [Organic Raspberries, Organic Hass Avocado]|
+----------------------------------------------------------+

+------------------------+-------------------+
|              consequent|         confidence|
+------------------------+-------------------+
|[Bag of Organic Bananas]|  0.521099116781158|
|                [Banana]| 0.4643478260869565|
|[Bag of Organic Bananas]| 0.4613385315139701|
|[Bag of Organic Bananas]| 0.4519846350832266|
|[Bag of Organic Bananas]| 0.4505169867060561|
|[Bag of Organic Bananas]| 0.4404332129963899|
|[Bag of Organic Bananas]| 0.4338461538461538|
|[Bag of Organic Bananas]|0.42270861833105333|
|                [Banana]| 0.4187779433681073|
|  [Organic Strawberries]|  0.414985590778098|
|[Bag of Organic Bananas]| 0.4108527131782946|
|[Bag of Organic Bananas]|0.40930232558139534|
|[Bag of Organic Bananas]|0.40706806282722513|
|[Bag of Organic Bananas]|0.39696485623003197|
|[Bag of Organic Bananas]| 0.3914780292942743|
|[Bag of Organic Bananas]|0.38862275449101796|
|                [Banana]| 0.3860811930405965|
|                [Banana]|0.38373305526590196|
|           [Large Lemon]| 0.3751205400192864|
|  [Organic Strawberries]|0.37389597644749756|
+------------------------+-------------------+
only showing top 20 rows
```

According to the output, customers who bought organic raspberries, organic avocados, and organic strawberries are also more likely to buy organic bananas. As you can see, bananas are a very popular item. This kind of lists could be the basis for "customers who bought this item also bought"-type recommendations.

Note In addition to FP-Growth, Spark MLlib includes another implementation of an algorithm for frequency pattern matching called PrefixSpan. While FP-Growth is indifferent about how the itemsets are ordered, PrefixSpan uses sequences, or ordered lists of itemsets, to discover sequential patterns in the dataset. PrefixSpan falls under a subgroup of algorithm known as sequential pattern mining. PrefixSpan is described in the paper "*Mining Sequential Patterns by Pattern-Growth: The PrefixSpan Approach*" by Jian Pei et al.

Content-Based Filtering

Content-based filtering provides recommendations by comparing information about an item, such as item name, description, or category, to a user's profile. Let's take, for instance, a content-based recommender system for movies. If the system determines that a user, based on his profile, has preference for Cary Grant movies, it might start recommending his movies such as *North by Northwest, To Catch a Thief,* and *An Affair to Remember*. The recommender might suggest movies by actors in the same genre such as Jimmy Stewart, Gregory Peck, or Clark Gable (classic movies). Movies directed by Alfred Hitchcock and George Cukor, frequent Cary Grant collaborators, might also be recommended. Despite its simplicity, content-based recommendation engines generally provide relevant results. It is also easy to implement. The system can provide recommendations immediately without waiting for explicit or implicit feedback from users, an arduous requirement that bedevil other methods such as collaborative filtering.

On the downside, content-based filtering lacks diversity and novelty in its recommendations. Viewers will sometimes want a broader selection of movies or something a little bit more avant-garde which might not be a perfect match to the viewer's profile. Scalability is another challenge that besets content-based recommender systems. To generate highly relevant recommendations, content-based engines require large amount of relevant domain-specific information about the item it's recommending.[xix] It is not enough for a movie recommender to base its suggestions just on the title, description, or genre. In-house data might have to be augmented with data from third-party sources such as IMDB or Rotten Tomatoes. Unsupervised learning methods such as Latent Dirichlet Allocation (LDA) can be used to create new topics out of metadata extracted from these data sources. I discuss LDA in detail in Chapter 4.

Netflix is leading in this area by creating thousands of *microgenres* that provide highly targeted personalized recommendations. For instance, not only does Netflix know that my wife loves watching Korean movies, it knows that she likes *Romantic Musical Korean Movies, Murder Mystery Zombie Korean Movies, Docudrama Gangster Korean Zombie Movies,* and her personal favorite *Courtroom Drama Legal Thriller Zombie Korean Movies.* Through a mechanical turk-like system, Netflix hired part-time movie buffs to manually assign descriptions and categories to thousands of movies and TV shows in its library. At last count, Netflix had 76,897 unique microgenres. This is a new level of personalization not seen anywhere outside of Netflix.

Spark MLlib does not include an algorithm for content-based filtering although Spark has the necessary components to help you develop your own implementation. To get you started, I suggest you look into a highly scalable similarity algorithm available in Spark known as "*Dimension Independent Matrix Square using MapReduce*" or DIMSUM for short. Check out `https://bit.ly/2YV6qTr` to learn more about DIMSUM.[xx]

Summary

Each method has its own strengths and weaknesses. In real-world scenarios, it is common practice to build a hybrid recommendation engine, combining multiple techniques to enhance results. Recommenders are fertile area for research. Considering the amount of revenue it generates for some of the world's largest companies, expect more developments in this area soon. The FP-Growth example was adapted from Bhavin Kukadia and Denny Lee's work at Databricks.[xxi]

References

i. René Girard; "René Girard and Mimetic Theory," imitatio.org, 2019, `www.imitatio.org/brief-intro`

ii. Michael Osborne; "How Retail Brands Can Compete And Win Using Amazon's Tactics," forbes.com, 2017, `www.forbes.com/sites/forbesagencycouncil/2017/12/21/how-retail-brands-can-compete-and-win-using-amazons-tactics/#4f4e55bc5e18`

iii. Ian MacKenzie et al.; "How retailers can keep up with consumers," mckinsey.com, 2013, www.mckinsey.com/industries/retail/our-insights/how-retailers-can-keep-up-with-consumers

iv. Nathan McAlone; "Why Netflix thinks its personalized recommendation engine is worth $1 billion per year," businessinsider.com, 2016, www.businessinsider.com/netflix-recommendation-engine-worth-1-billion-per-year-2016-6

v. Bernard Marr; "The Amazing Ways Chinese Tech Giant Alibaba Uses Artificial Intelligence And Machine Learning," 2018, www.forbes.com/sites/bernardmarr/2018/07/23/the-amazing-ways-chinese-tech-giant-alibaba-uses-artificial-intelligence-and-machine-learning/#686ffa0117a9

vi. William Vorhies; "5 Types of Recommenders," datasciencecentral.com, 2017, www.datasciencecentral.com/m/blogpost?id=6448529%3ABlogPost%3A512183

vii. Carol McDonald; "Building a Recommendation Engine with Spark," mapr.com, 2015, https://mapr.com/ebooks/spark/08-recommendation-engine-spark.html

viii. Burak Yavuz et al.; "Scalable Collaborative Filtering with Apache Spark MLlib," Databricks.com, 2014, https://databricks.com/blog/2014/07/23/scalable-collaborative-filtering-with-spark-mllib.html

ix. IBM; "Introduction to Apache Spark lab, part 3: machine learning," IBM.com, 2017, https://dataplatform.cloud.ibm.com/exchange/public/entry/view/5ad1c820f57809ddec9a040e37b4af08

x. Spark; "Class ALS," spark.apache.org, 2019, https://spark.apache.org/docs/2.0.0/api/java/org/apache/spark/mllib/recommendation/ALS.html

xi. Robert M. Bell and Yehuda Koren; "Scalable Collaborative Filtering with Jointly Derived Neighborhood Interpolation Weights," acm.org, 2007, https://dl.acm.org/citation.cfm?id=1442050

xii. Apache Spark; "Collaborative Filtering," spark.apache.org, 2019, https://spark.apache.org/docs/latest/ml-collaborative-filtering.html

xiii. Apache Spark; "Collaborative Filtering," spark.apache.org, 2019, https://spark.apache.org/docs/latest/ml-collaborative-filtering.html

xiv. Margaret Rouse; "association rules (in data mining)," techtarget.com, 2018, https://searchbusinessanalytics.techtarget.com/definition/association-rules-in-data-mining

xv. Apache Spark; "Frequent Pattern Mining," spark.apache.org, 2019, https://spark.apache.org/docs/2.3.0/mllib-frequent-pattern-mining.html

xvi. Jiawei Han et. al.; "Mining frequent patterns without candidate generation," acm.org, 2000, https://dl.acm.org/citation.cfm?doid=335191.335372

xvii. Jiawei Han, et al.; "Mining frequent patterns without candidate generation," acm.org, 2000, https://dl.acm.org/citation.cfm?id=335372%C3%DC

xviii. Bhavin Kukadia and Denny Lee; "Simplify Market Basket Analysis using FP-growth on Databricks," Databricks.com, 2018, https://databricks.com/blog/2018/09/18/simplify-market-basket-analysis-using-fp-growth-on-databricks.html

xix. Tyler Keenan; "What Is Content-Based Filtering?", upwork.com, 2019, www.upwork.com/hiring/data/what-is-content-based-filtering/

xx. Reza Zadeh; "Efficient similarity algorithm now in Apache Spark, thanks to Twitter," Databricks.com, 2014, https://databricks.com/blog/2014/10/20/efficient-similarity-algorithm-now-in-spark-twitter.html

xxi. Bhavin Kukadia and Denny Lee; "Simplify Market Basket Analysis using FP-growth on Databricks," Databricks.com, 2018, https://databricks.com/blog/2018/09/18/simplify-market-basket-analysis-using-fp-growth-on-databricks.html

CHAPTER 6

Graph Analysis

Tell me where I'm going to die so I don't go there.

—Charlie Munger[i]

Graph analysis is a data analytic technique for determining the strength and direction of relationships between objects in a graph. Graphs are mathematical structures used to model relationships and processes between objects.[ii] They can be used to represent complex relationships and dependencies in your data. Graphs are composed of vertices or nodes that represent the entities in the system. These vertices are connected by edges which represent the relationships between those entities.[iii]

Introduction to Graphs

It may not be immediately obvious, but graphs are everywhere. Social networks such as LinkedIn, Facebook, and Twitter are graphs. The Internet and the World Wide Web are graphs. Computer networks are graphs. Water utility pipelines are graphs. Road networks are graphs. Graph processing frameworks such as GraphX have graph algorithms and operators specifically designed to deal with graph-oriented data.

Undirected Graph

Undirected graphs are graphs that have edges with no direction. The edges in an undirected graph can be traversed in both directions and represent a two-way relationship. Figure 6-1 shows an undirected graph with three nodes and three edges.[iv]

B. Quinto, *Next-Generation Machine Learning with Spark*, https://doi.org/10.1007/978-1-4842-5669-5_6

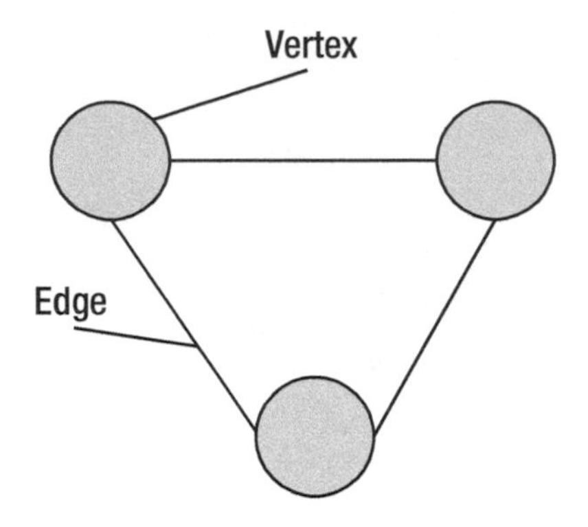

Figure 6-1. *Undirected graph*

Directed Graph

Directed graphs have edges with direction that indicate a one-way relationship. In directed graphs, each edge can only be traversed in one direction (Figure 6-2).

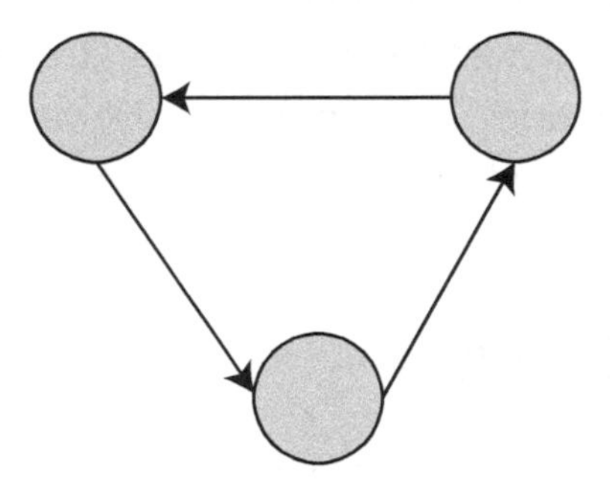

Figure 6-2. *Directed graph*

Directed Multigraph

A multigraph has multiple edges between nodes (Figure 6-3).

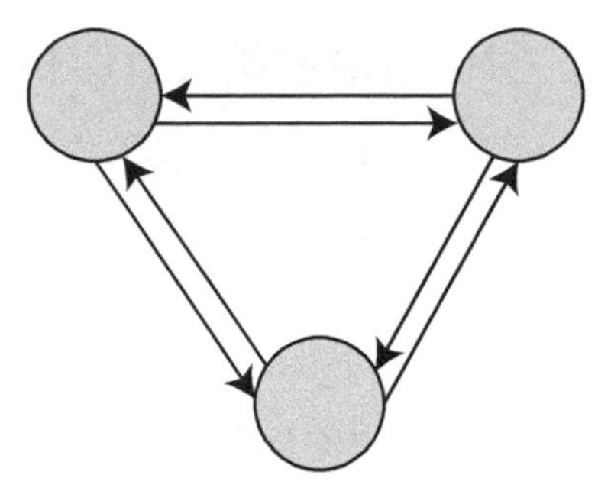

Figure 6-3. *Directed multigraph*

Property Graph

A property graph is a directed multigraph where the vertices and edges have user-defined attributes[v] (Figure 6-4).

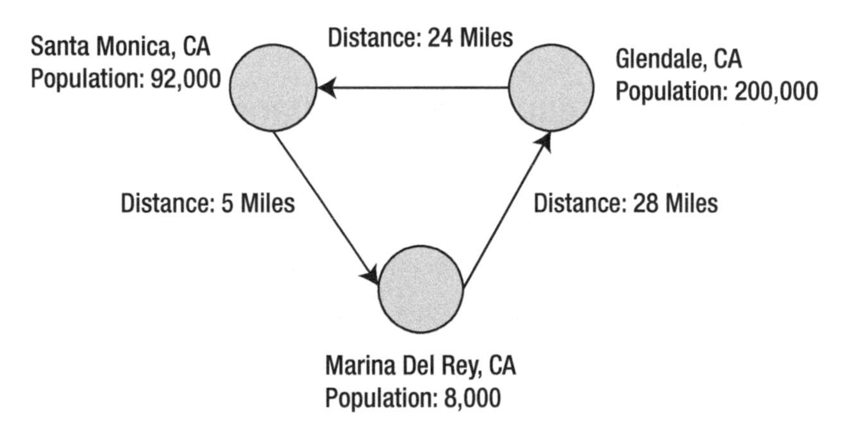

Figure 6-4. *Property graph*

Graph Analysis Use Cases

Graph analysis has flourished across a wide range of industries. Any project that utilizes high volumes of connected data is a perfect use case for graph analysis. This is not meant to be a comprehensive list of use cases, but it should give you an idea of the possibilities with graph analytics.

Fraud Detection and Anti-Money Laundering (AML)

Fraud detection and anti-money laundering is perhaps one of the most well-publicized use cases for graph analytics. Manually sifting through millions of bank transactions to identify fraud and money laundering activities is daunting, if not impossible. Sophisticated methods to hide fraud through a complex series of interconnected and seemingly harmless transactions have compounded the problem. Traditional techniques to detect and prevent these types of attacks have been rendered obsolete in this day in age. Graph analysis is the perfect solution to this problem by making it easy to connect suspicious transactions with other abnormal behavior patterns. High-performance graph processing APIs like GraphX allow for complex traversals from one transaction to another very rapidly. Perhaps the most well-known example of this is the Panama Papers scandal. By identifying connections between millions of leaked documents using graph analysis, analysts were able to uncover hidden assets in offshore bank accounts owned by high-profile foreign leaders, politicians, and even the Queen.[vi]

Data Governance and Regulatory Compliance

A typical large organization stores millions of data files in a central data repository such as a data lake. This requires proper master data management which means tracking data lineage every time a file is modified or a new copy is created. By tracking data lineage, organizations can follow the movement of data from source to destination, providing visibility into all the ways it has changed from point to point.[vii] A sound data lineage policy promotes confidence in the data and enables accurate business decisions. Data lineage is also a requirement in maintaining data-related regulatory compliance such as the General Data Protection Regulation (GDPR). Getting caught violating GDPR can mean a hefty fine. Graph analysis is perfect for these use cases.

Risk Management

To manage risk, financial institutions such as hedge funds and investment banks use graph analysis to identify interconnected risks and patterns that has the potential of setting off "black swan" events. Using the 2008 financial crisis as the perfect example, graph analysis could have provided visibility into the complex process of securitizing derivatives by shining a light on the intricate interdependencies between mortgage-backed securities (MBS), collateralized debt obligations (CDOs), and the credit default swaps that are placed on tranches of CDOs.

Transportation

The air traffic network is a graph. The airports represent the vertices, while the routes represent the edges. Commercial flight plans are created using graph analysis. Airlines and logistics companies use graph analysis for route optimization to determine the safest or fastest route possible. These optimizations ensure safety and can save time and money.

Social Networking

Social networking is the most intuitive use case for graph analysis. Almost everyone these days has a "social graph" - a graph that represents a person's social relationship with other people or group in a social network. If you're on Facebook, Twitter, Instagram, or LinkedIn, you have a social graph. A good example is Facebook's use of Apache Giraph, another open source graph framework, to process and analyze one trillion edges for content ranking and recommendations.[viii]

Network Infrastructure Management

The Internet is a giant graph of interconnected routers, servers, desktops, IoT, and mobile devices. Government and corporate networks can be regarded as subgraphs of the Internet. Even small home networks are graphs. The devices represent the vertices and the network connections represent the edges. Graph analytics provides network administrators the ability to visualize complex network topologies, which is useful for monitoring and troubleshooting purposes. This is particularly valuable for large corporate networks that consist of thousands of connected devices.

Introduction to GraphX

GraphX is Spark's RDD-based graph processing API. In addition to graph operators and algorithms, GraphX provides data types for storing graph data.[ix]

Graph

The property graph is represented by an instance of the Graph class. Just like RDDs, property graphs are partitioned and distributed across executors and can be recreated in the event of a crash. Property graphs are immutable, which means creating a new graph is necessary to change the structure or values of the graph.[x]

VertexRDD

The vertices in a property graph are represented by a VertexRDD. Only one entry for each vertex is stored in a VertexRDD.

Edge

The Edge class contains the source and destination ids corresponding to the source and destination vertex identifiers as well as the attribute that stores the edge property.[xi]

EdgeRDD

The edges in a property graph are represented by an EdgeRDD.

EdgeTriplet

A combination of an edge and the two vertices that it connects is represented by an instance of an EdgeTriplet class. It also contains the attributes of the edge and the vertices that it connects.

EdgeContext

The EdgeContext class exposes the triplet fields as well as the functions to send messages to the source and destination vertex.[xii]

GraphX Example

For this example (Listing 6-1), I will use GraphX to analyze distances between different cities in Southern California. We will construct a property graph that looks like Figure 6-5 based on the data in Table 6-1.

Table 6-1. Distances Between Different Southern California Cities

Source	Destination	Distance
Santa Monica, CA	Marina Del Rey, CA	5 Miles
Santa Monica, CA	Glendale, CA	24 Miles
Marina Del Rey, CA	Glendale, CA	28 Miles
Glendale, CA	Pasadena, CA	9 Miles
Pasadena, CA	Glendale, CA	9 Miles

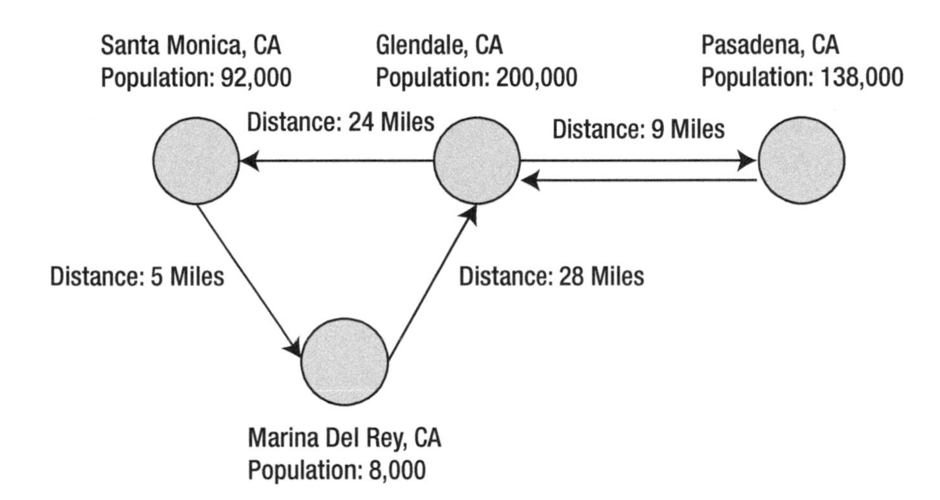

Figure 6-5. *Property graph*

Listing 6-1. GraphX example[xiii]

```
importorg.apache.spark.rdd.RDD
importorg.apache.spark.graphx._

// Define the vertices.

val vertices = Array((1L, ("Santa Monica","CA")),(2L,("Marina Del
Rey","CA")),(3L, ("Glendale","CA")),(4L, ("Pasadena","CA")))

valvRDD = sc.parallelize(vertices)
```

```
// Define the edges.

val edges = Array(Edge(1L,2L,5),Edge(1L,3L,24),Edge(2L,3L,28),Edge(3L,4L,9),
Edge(4L,3L,9))

val eRDD = sc.parallelize(edges)

// Create a property graph.

val graph = Graph(vRDD,eRDD)

graph.vertices.collect.foreach(println)

(3,(Glendale,CA))
(4,(Pasadena,CA))
(1,(Santa Monica,CA))
(2,(Marina Del Rey,CA))

graph.edges.collect.foreach(println)
Edge(1,2,5)
Edge(1,3,24)
Edge(2,3,28)
Edge(3,4,9)
Edge(4,3,9)

//Return the number of vertices.

val numCities = graph.numVertices
numCities: Long = 4

// Return the number of edges.

val numRoutes = graph.numEdges
numRoutes: Long = 5

// Return the number of ingoing edges for each vertex.

graph.inDegrees.collect.foreach(println)
(3,3)
(4,1)
(2,1)
```

```
// Return the number of outgoing edges for each vertex.

graph.outDegrees.collect.foreach(println)
(3,1)
(4,1)
(1,2)
(2,1)

// Return all routes that is less than 20 miles.

graph.edges.filter{ case Edge(src, dst, prop) => prop < 20 }.collect.
foreach(println)

Edge(1,2,5)
Edge(3,4,9)
Edge(4,3,9)

// The EdgeTriplet class includes the source and destination attributes.

graph.triplets.collect.foreach(println)

((1,(Santa Monica,CA)),(2,(Marina Del Rey,CA)),5)
((1,(Santa Monica,CA)),(3,(Glendale,CA)),24)
((2,(Marina Del Rey,CA)),(3,(Glendale,CA)),28)
((3,(Glendale,CA)),(4,(Pasadena,CA)),9)
((4,(Pasadena,CA)),(3,(Glendale,CA)),9)

// Sort by farthest route.

graph.triplets.sortBy(_.attr, ascending=false).collect.foreach(println)
((2,(Marina Del Rey,CA)),(3,(Glendale,CA)),28)
((1,(Santa Monica,CA)),(3,(Glendale,CA)),24)
((3,(Glendale,CA)),(4,(Pasadena,CA)),9)
((4,(Pasadena,CA)),(3,(Glendale,CA)),9)
((1,(Santa Monica,CA)),(2,(Marina Del Rey,CA)),5)

// mapVertices applies a user-specified function to every vertex.

val newGraph = graph.mapVertices((vertexID,state) => "TX")
```

```
newGraph.vertices.collect.foreach(println)
(3,TX)
(4,TX)
(1,TX)
(2,TX)
```

```
// mapEdges applies a user-specified function to every edge.

val newGraph = graph.mapEdges((edge) => "500")

Edge(1,2,500)
Edge(1,3,500)
Edge(2,3,500)
Edge(3,4,500)
Edge(4,3,500)
```

Graph Algorithms

GraphX comes with built-in implementation of several common graph algorithms such as PageRank, triangle count, and connected components.

PageRank

PageRank is an algorithm that was originally developed by Google to determine the importance of web pages. A web page with a higher PageRank is more relevant than a page with a lower PageRank. The PageRank of a page depends on the PageRank of the pages linking to it; thus, it is an iterative algorithm. The number of high-quality links also contributes to the PageRank of a page. GraphX includes a built-in implementation of PageRank. GraphX comes with a static and dynamic versions of PageRank.

Dynamic PageRank

Dynamic PageRank executes until the ranks stop updating by more than the specified tolerance (i.e., until the ranks converge).

```
val dynamicPageRanks = graph.pageRank(0.001).vertices

val sortedRanks = dynamicPageRanks.sortBy(_._2,ascending=false)

sortedRanks.collect.foreach(println)
```

```
(3,1.8845795504535865)
(4,1.7507334787248419)
(2,0.21430059110133595)
(1,0.15038637972023575)
```

Static PageRank

Static PageRank executes for a set number of times.

```
val staticPageRanks = graph.staticPageRank(10)

val sortedRanks = staticPageRanks.vertices.sortBy(_._2,ascending=false)

sortedRanks.collect.foreach(println)
(4,1.8422463479403317)
(3,1.7940036520596683)
(2,0.21375000000000008)
(1,0.15000000000000005)
```

TriangleCount

A triangle consists of three connected vertices. The triangle count algorithm provides a measure of clustering by determining the number of triangles passing through each vertex. In Figure 6-5, Santa Monica, Marina Del Rey, and Glendale are all part of a triangle, while Pasadena is not.

```
val triangleCount = graph.triangleCount()

triangleCount.vertices.collect.foreach(println)
(3,1)
(4,0)
(1,1)
(2,1)
```

ConnectedComponents

The connected components algorithm determines the membership of each vertex in a subgraph. The algorithm returns the vertex id of the lowest-numbered vertex in a subgraph as the vertex's property. Figure 6-6 shows two connected components. I show an example in Listing 6-2.

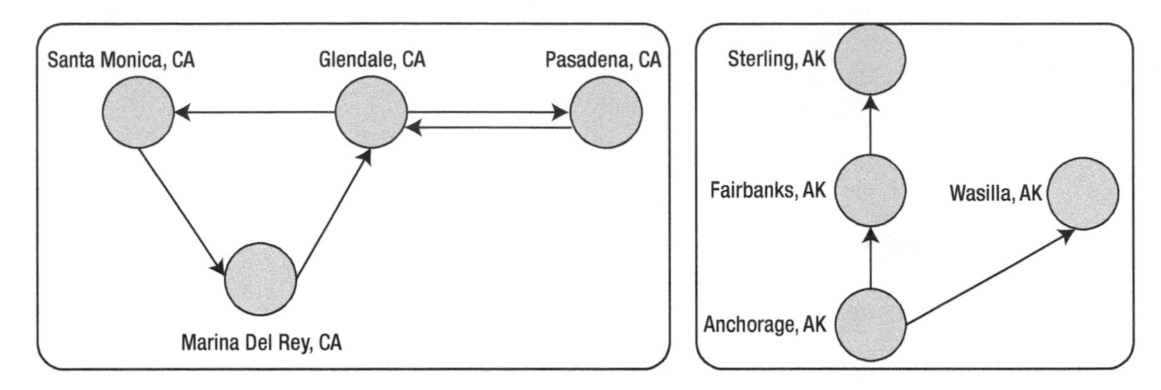

Figure 6-6. *Connected components*

Listing 6-2. Determine the Connected Components

```
val vertices = Array((1L, ("Santa Monica","CA")),(2L,("Marina Del Rey","CA")),
(3L, ("Glendale","CA")),(4L, ("Pasadena","CA")),(5L, ("Anchorage","AK")),
(6L, ("Fairbanks","AK")),(7L, ("Sterling","AK")),(8L, ("Wasilla","AK")))

val vRDD = sc.parallelize(vertices)

val edges = Array(Edge(1L,2L,5),Edge(1L,3L,24),Edge(2L,3L,28),Edge(3L,4L,9),
Edge(4L,3L,9),Edge(5L,6L,32),Edge(6L,7L,28),Edge(5L,8L,17))

val eRDD = sc.parallelize(edges)

val graph = Graph(vRDD,eRDD)

graph.vertices.collect.foreach(println)

(6,(Fairbanks,AK))
(3,(Glendale,CA))
(4,(Pasadena,CA))
(1,(Santa Monica,CA))
(7,(Sterling,AK))
(8,(Wasilla,AK))
(5,(Anchorage,AK))
(2,(Marina Del Rey,CA))
```

```
graph.edges.collect.foreach(println)

Edge(1,2,5)
Edge(1,3,24)
Edge(2,3,28)
Edge(3,4,9)
Edge(4,3,9)
Edge(5,6,32)
Edge(5,8,17)
Edge(6,7,28)

val connectedComponents = graph.connectedComponents()

connectedComponents.vertices.collect.foreach(println)

(6,5)
(3,1)
(4,1)
(1,1)
(7,5)
(8,5)
(5,5)
(2,1)
```

GraphFrames

GraphFrames is a graph processing library built on top of DataFrames. At the time of this writing, GraphFrames is still in active development, but it is just a matter of time before it becomes part of the core Apache Spark framework. There are a few things that make GraphFrames more powerful than GraphX. All GraphFrames algorithms are available in Java, Python, and Scala. GraphFrames can be accessed using the familiar DataFrames API and Spark SQL. It also fully supports DataFrame data sources which allow reading and writing graphs using several supported formats and data sources such as relational data sources, CSV, JSON, and Parquet.[xiv] Listing 6-3 shows an example on how to use GraphFrames.[xv]

Listing 6-3. GraphFrames Example

```
spark-shell --packages graphframes:graphframes:0.7.0-spark2.4-s_2.11

import org.graphframes._

val vDF = spark.createDataFrame(Array((1L, "Santa Monica","CA"),
(2L,"Marina Del Rey","CA"),(3L, "Glendale","CA"),(4L, "Pasadena","CA"),
(5L, "Anchorage","AK"),(6L, "Fairbanks","AK"),(7L, "Sterling","AK"),
(8L, "Wasilla","AK"))).toDF("id","city","state")

vDF.show
+---+--------------+-----+
| id|          city|state|
+---+--------------+-----+
|  1|  Santa Monica|   CA|
|  2|Marina Del Rey|   CA|
|  3|      Glendale|   CA|
|  4|      Pasadena|   CA|
|  5|     Anchorage|   AK|
|  6|     Fairbanks|   AK|
|  7|      Sterling|   AK|
|  8|       Wasilla|   AK|
+---+--------------+-----+

val eDF = spark.createDataFrame(Array((1L,2L,5),(1L,3L,24),(2L,
3L,28),(3L,4L,9),(4L,3L,9),(5L,6L,32),(6L,7L,28),(5L,8L,17))).
toDF("src","dst","distance")

eDF.show
+---+----+--------+
|src|dest|distance|
+---+----+--------+
|  1|   2|       5|
|  1|   3|      24|
|  2|   3|      28|
|  3|   4|       9|
|  4|   3|       9|
```

```
|  5|   6|       32|
|  6|   7|       28|
|  5|   8|       17|
+---+----+--------+

val graph = GraphFrame(vDF,eDF)

graph.vertices.show
+---+-------------+-----+
| id|         city|state|
+---+-------------+-----+
|  1| Santa Monica|   CA|
|  2|Marina Del Rey|   CA|
|  3|     Glendale|   CA|
|  4|     Pasadena|   CA|
|  5|    Anchorage|   AK|
|  6|    Fairbanks|   AK|
|  7|     Sterling|   AK|
|  8|      Wasilla|   AK|
+---+-------------+-----+

graph.edges.show

+---+---+--------+
|src|dst|distance|
+---+---+--------+
|  1|  2|       5|
|  1|  3|      24|
|  2|  3|      28|
|  3|  4|       9|
|  4|  3|       9|
|  5|  6|      32|
|  6|  7|      28|
|  5|  8|      17|
+---+---+--------+
```

```
graph.triplets.show

+-------------------+---------+-------------------+
|                src|     edge|                dst|
+-------------------+---------+-------------------+
|[1, Santa Monica,...|[1, 3, 24]|   [3, Glendale, CA]|
|[1, Santa Monica,...| [1, 2, 5]|[2, Marina Del Re...|
|[2, Marina Del Re...|[2, 3, 28]|   [3, Glendale, CA]|
|   [3, Glendale, CA]| [3, 4, 9]|   [4, Pasadena, CA]|
|   [4, Pasadena, CA]| [4, 3, 9]|   [3, Glendale, CA]|
|  [5, Anchorage, AK]|[5, 8, 17]|    [8, Wasilla, AK]|
|  [5, Anchorage, AK]|[5, 6, 32]|  [6, Fairbanks, AK]|
|  [6, Fairbanks, AK]|[6, 7, 28]|   [7, Sterling, AK]|
+-------------------+---------+-------------------+

graph.inDegrees.show

+---+--------+
| id|inDegree|
+---+--------+
|  7|       1|
|  6|       1|
|  3|       3|
|  8|       1|
|  2|       1|
|  4|       1|
+---+--------+

graph.outDegrees.show

+---+---------+
| id|outDegree|
+---+---------+
|  6|        1|
|  5|        2|
|  1|        2|
```

```
|  3|        1|
|  2|        1|
|  4|        1|
+---+---------+

sc.setCheckpointDir("/tmp")

val connectedComponents = graph.connectedComponents.run

connectedComponents.show

+---+-------------+-----+---------+
| id|         city|state|component|
+---+-------------+-----+---------+
|  1| Santa Monica|   CA|        1|
|  2|Marina Del Rey|   CA|        1|
|  3|     Glendale|   CA|        1|
|  4|     Pasadena|   CA|        1|
|  5|    Anchorage|   AK|        5|
|  6|    Fairbanks|   AK|        5|
|  7|     Sterling|   AK|        5|
|  8|      Wasilla|   AK|        5|
+---+-------------+-----+---------+
```

Summary

This chapter serves as an introduction to graph analysis using GraphX and GraphFrames. Graph analysis is an exciting and rapidly growing area of study with wide and far-reaching applications. We just scratched the surface of what GraphX and GraphFrames can do. While it may not fit every use case, graph analysis is great for what it was designed to do and is an indispensable addition to your analytic toolset.

References

i. Michael Simmons; "What Self-Made Billionaire Charlie Munger Does Differently," inc.com, 2019, www.inc.com/michael-simmons/what-self-made-billionaire-charlie-munger-does-differently.html

ii. Carol McDonald; "How to Get Started Using Apache Spark GraphX with Scala," mapr.com, 2015, https://mapr.com/blog/how-get-started-using-apache-spark-graphx-scala/

iii. Nvidia; "Graph Analytics," developer.nvidia.com, 2019, https://developer.nvidia.com/discover/graph-analytics

iv. MathWorks; "Directed and Undirected Graphs," mathworks.com, 2019, www.mathworks.com/help/matlab/math/directed-and-undirected-graphs.html

v. Rishi Yadav; "Chapter 11, Graph Processing using GraphX and GraphFrames," Packt Publishing, 2017, Apache Spark 2.x Cookbook

vi. Walker Rowe; "Use Cases for Graph Databases," bmc.com, 2019, www.bmc.com/blogs/graph-database-use-cases/

vii. Rob Perry; "Data Lineage: The Key to Total GDPR Compliance," insidebigdata.com, 2018, https://insidebigdata.com/2018/01/29/data-lineage-key-total-gdpr-compliance/

viii. Avery Ching et al.; "One Trillion Edges: Graph Processing at Facebook-Scale," research.fb.com, 2015, https://research.fb.com/publications/one-trillion-edges-graph-processing-at-facebook-scale/

ix. Mohammed Guller; "Graph Processing with Spark," Apress, 2015, Big Data Analytics with Spark

x. Spark; "The Property Graph," spark.apache.org, 2015, https://spark.apache.org/docs/latest/graphx-programming-guide.html#the-property-graph

xi. Spark; "Example Property Graph," spark.apache.org, 2019, https://spark.apache.org/docs/latest/graphx-programming-guide.html#the-property-graph

xii. Spark; "Map Reduce Triplets Transition Guide," spark.apache.org, 2019, https://spark.apache.org/docs/latest/graphx-programming-guide.html#the-property-graph

xiii. Carol McDonald; "How to Get Started Using Apache Spark GraphX with Scala," mapr.com, 2015, https://mapr.com/blog/how-get-started-using-apache-spark-graphx-scala/

xiv. Ankur Dave et al.; "Introducing GraphFrames," Databricks.com, 2016, https://databricks.com/blog/2016/03/03/introducing-graphframes.html

xv. Rishi Yadav; "Chapter 11, Graph Processing Using GraphX and GraphFrames," Packt Publishing, 2017, Apache Spark 2.x Cookbook

CHAPTER 7

Deep Learning

Minsky and Papert would have contributed more if they had produced a solution to this problem rather than beating the Perceptron to death.

—Francis Crick[i]

Deep learning is a subfield of machine learning and artificial intelligence that uses deep, multilayered artificial neural networks (Figure 7-1). It is all the rage these days due to its ability to perform human-like tasks with exceptional accuracy. Advances in the field of deep learning have opened up exciting possibilities in the areas of computer vision, speech recognition, gaming, and anomaly detection, to mention a few. The mainstream availability of GPUs (Graphics Processing Unit) catapulted the global adoption of artificial intelligence almost overnight. High-performance computing has gotten powerful enough that tasks that seemed computationally impossible just a few years ago are now routinely performed on multi-GPU cloud instances or clusters of inexpensive machines. This has allowed numerous innovations in the field of artificial intelligence to develop at a rate that was not possible in the past.

B. Quinto, *Next-Generation Machine Learning with Spark*, https://doi.org/10.1007/978-1-4842-5669-5_7

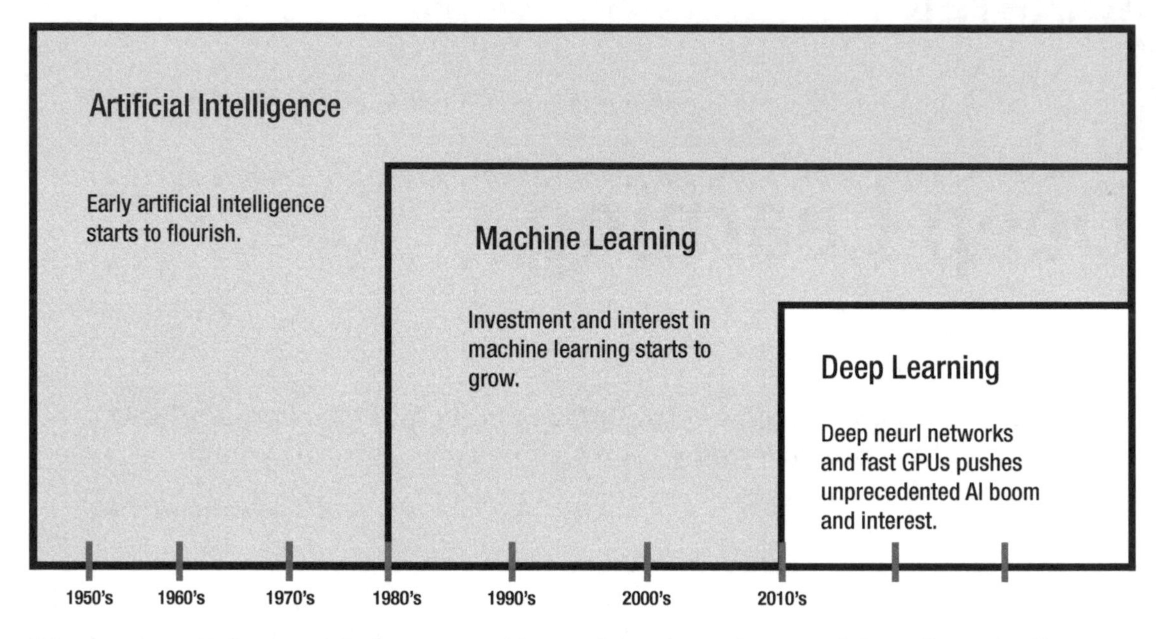

Figure 7-1. *Relationship between AI, machine learning, and deep learning*[ii]

Deep learning is responsible for many of the recent breakthroughs in artificial intelligence. While deep learning can be used for more mundane classification tasks, its true power shines when applied to more complex problems such as medical diagnosis, facial recognition, self-driving cars, fraud analysis, and intelligent voice-controlled assistants.[iii] In certain areas, deep learning has enabled computers to match and even outperform human abilities. For instance, with deep learning, Google DeepMind's AlphaGo program was able to defeat South Korean Master Lee Se-dol in Go, one of the most complex board game ever invented. Go is infinitely more complicated than other board games such as chess. It has a 10 to the power of 170 possible board configurations which is more than the number of atoms in the universe.[iv] Another great example is Tesla. Tesla uses deep learning[v] to power its autopilot features, processing data from several surround cameras, ultrasonic sensors, and forward-facing radar built in every car it manufactures. To deliver high-performance processing capabilities to its deep neural networks, Tesla fits its self-driving cars with a computer using its own AI chip, stocked with a GPU, CPU, and deep learning accelerators that can deliver 144 trillion operations per second (TOPS).[vi]

Similar to other machine learning algorithms, deep neural networks generally perform better with more training data. This is where Spark comes into the picture. But while Spark MLlib includes an extensive array of machine learning algorithms, its deep learning support is still limited at the time of this writing (Spark MLlib includes a multilayer perceptron classifier, a fully connected, feedforward neural network limited to training shallow networks and performing transfer learning). However, thanks to the community of developers and companies such as Google and Facebook, there are several third-party distributed deep learning libraries and frameworks that integrate with Spark. We'll explore some of the most popular ones. For this chapter I will focus on one of the most popular deep learning frameworks: Keras. For distributed deep learning, we will use Elephas and Distributed Keras (Dist-Keras). We will use Python throughout this chapter. We will use PySpark for our distributed deep learning examples.

Neural Networks

Neural networks are a class of algorithms that operate like interconnected neurons of the human brain. A neural network contains multiple layers that consist of interconnected nodes. There is usually an input layer, one or more hidden layers, and an output layer (Figure 7-2). Data goes through the neural network via the input layer. The hidden layers process the data through a network of weighted connections. The nodes in the hidden layer assign weights to the input and combine it with a set of coefficients along the way. The data goes through a node's activation function which determines the output of the layer. Finally, the data reaches the output layer which produces the final output of the neural network.[vii]

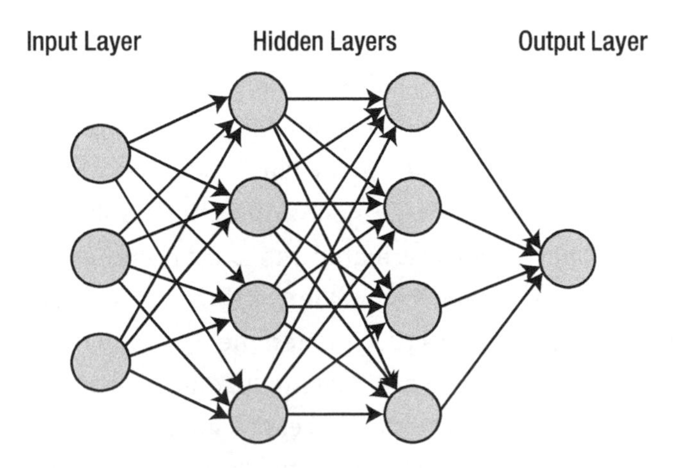

Figure 7-2. *A simple neural network*

Neural networks with several hidden layers are known as "deep" neural networks. The more layers, the deeper the network, and generally the deeper the network, the more sophisticated the learning become and the more complex problems it can solve. This is what gives deep neural networks the ability to achieve state-of-the-art accuracy. There are several types of neural networks: feedforward neural networks, recurrent neural networks, and autoencoder neural networks, each with its own capabilities and designed for different purposes. Feedforward neural network or multilayer perceptron works well with structured data. Recurrent neural networks are a good option for sequential data such as audio, time-series, or text.[viii] Autoencoders are a good choice for anomaly detection and dimensionality reduction. One type of neural network, the convolutional neural network, has taken the world by storm lately due to its revolutionary performance in the field of computer vision.

A Brief History of Neural Networks

The first neural network, the McCulloch-Pitts Neuron, was developed in 1943 by Warren McCulloch and Walter Pitts and described in their seminal paper "A Logical Calculus of the Idea Immanent in Nervous Activity."[ix] The McCulloch-Pitts Neuron doesn't resemble the modern neural network, but it was the seed that paved the way for the birth of artificial intelligence as we know it today. The Perceptron, introduced in 1958 by Frank Rosenblatt, was the next revolutionary development in the field of neural networks. The Rosenblatt Perceptron is a simple single-layer neural network used for binary classification. It was conceived at the Cornell Aeronautical Laboratory, initially simulated in software using an IBM 704. Rosenblatt's work culminated in the creation of the Mark I Perceptron, a custom-built machine designed for image recognition. Interest in the Perceptron began to wane in the late 1960s after Marvin Minsky and Seymour Papert published *Perceptrons*, a highly influential book that expounded on the Perceptron's limitations, particularly its inability to learn nonlinear functions.[x] Minsky and Papert's overzealous and sometimes hostile (as they would later admit years later) repudiation[xi] of connectionist models precipitated the *AI winter* of the 1980s, a period of diminished interest and funding in artificial intelligence research. The situation was further exacerbated due in part by the success of support vector machines (SVM) and graphical models and the growing popularity of symbolic artificial intelligence, the prevalent AI paradigm from the 1950s until the end of the 1980s.[xii] Nevertheless, a number of scientists and researchers such as Jim Anderson, David Willshaw, and

Stephen Grossberg quietly continued their research on neural networks. In 1975, Kunihiko Fukushima developed the first multilayered neural network, the neocognitron, as a research scientist at the NHK Science & Technology Research Laboratories in Segataya, Tokyo, Japan.[xiii]

Neural networks had a resurgence in the mid-1980s during which many of the fundamental concepts and techniques central to modern-day deep learning were invented by a number of pioneers. In 1986, Geoffrey Hinton (considered by many as the "Godfather of AI") popularized backpropagation, the fundamental method by which neural networks learn and is central to almost all modern neural network implementations.[xiv] Equally important, Hinton and his team popularized the idea of using multilayer neural networks to learn nonlinear functions, directly addressing Minsky and Papert's criticisms in *Perceptrons.*[xv] Hinton's work was instrumental in bringing neural networks back from obscurity and into the spotlight once again. Yann LeCun, inspired by Fukushima's neocognitron, developed convolutional neural networks in the late 1980s. While at AT&T Bell Labs in 1988, Yann LeCun used convolutional neural networks to recognize handwritten characters. AT&T sold the system to banks to read and recognize handwriting in checks and is considered as one of the first real-world applications of neural networks.[xvi] Yoshua Bengio is known for his introduction of the concept of word embeddings and his more recent work with Ian Goodfellow on generative adversarial networks (GANs), a type of deep neural network architecture consisting of two neural networks opposed or "adversarial" with each other. Adversarial training has been described by Yann LeCun as "the most interesting idea in the last 10 years in machine learning."[xvii] The three pioneers recently received the Turing Award for 2018.

Nonetheless, despite these innovations practical use of neural networks remained out of reach to everyone, but the largest corporations and well-funded universities due to the significant amount computational resources needed to train them. It was the widespread availability of affordable, high-performance GPUs that pushed neural networks into the mainstream. Previously thought of as unscalable and impractical to train, neural networks have suddenly started delivering on the promise of human-like performance in the field of computer vision, speech recognition, gaming, and other historically hard machine learning tasks with exceptional accuracy using GPUs. The world took notice when AlexNet, a GPU-trained convolutional neural network by Alex Krizhevsky, Ilya Sutskever, and Geoff Hinton, won the ImageNet Large Scale Visual Recognition contest in 2012.[xviii] There were other GPU-trained CNN that previously won the competition, but it was AlexNet's record-breaking accuracy rate that got everyone's attention. It won the contest by such a large margin, achieving more 10.8 percentage points lower than the runner-up, a top-5 error of 15.3%.[xix] AlexNet also popularized most of the principles that are now common in most CNN architectures such as the use of rectified linear unit (ReLU) activation function, the use of dropouts and data augmentation to reduce overfitting, overlapping pooling, and training on multiple GPUs.

There is recently an unprecedented amount of innovation in the field of AI from both the academia and private sector. Companies such as Google, Amazon, Microsoft, Facebook, IBM, Tesla, Nvidia, and others are investing significant resources in deep learning, further pushing the boundaries of artificial intelligence. Leading universities such as Stanford University, Carnegie Mellon University, and MIT and research organizations such as OpenAI and the Allen Institute of Artificial Intelligence are publishing ground-breaking research at an astonishing rate. Globally, impressive innovations are coming out of countries like China, Canada, the United Kingdom, and Australia, to mention a few.

Convolutional Neural Networks

A convolutional neural network (convnet or CNN for short) is a type of neural network that is particularly good at analyzing images (although they can also be applied to audio and text data). The neurons in the layers of a convolutional neural network are arranged in three dimensions: height, width, and depth. CNNs use convolutional layers to learn local patterns in its input feature space (images) such as textures and edges. In contrast, fully connected (dense) layers learn global patterns.[xx] The neurons in a convolutional layer are only connected to a small region of the layer preceding it instead of all of the neurons as is the case with dense layers. A dense layer's fully connected structure can lead to extremely large number of parameters which is inefficient and could quickly lead to overfitting.[xxi]

Before we continue with our discussion of convolutional neural networks, it is important to understand the concept of color models. Images are represented as a collection of pixels.[xxii] A pixel is the smallest unit of information that makes up an image. Pixel values for grayscale images range from 0 (black) to 255 (white) as shown in Figure 7-3(a). Grayscale images only have one channel. RGB is the most traditional color model. RGB images have three channels: red, green, and blue. This means each pixel in an RGB image is represented by three 8-bit numbers (Figure 7-3(b)), each ranging from 0 to 255. Using different combinations of these channels displays a different color.

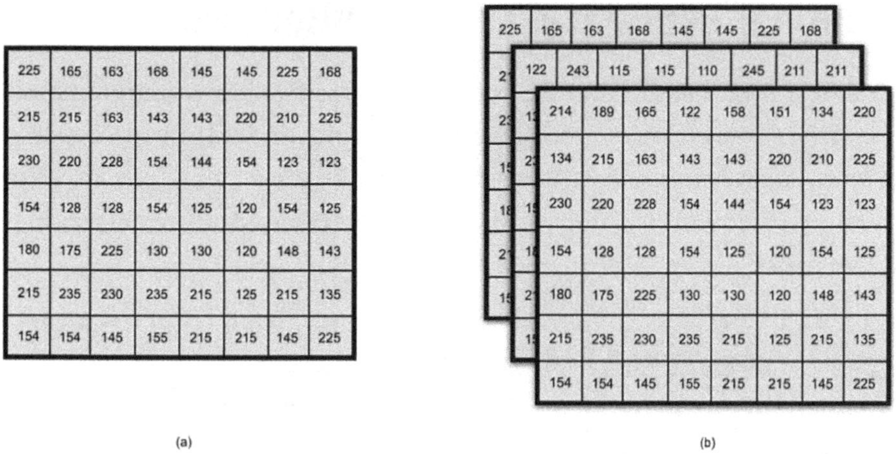

225	165	163	168	145	145	225	168
215	215	163	143	143	220	210	225
230	220	228	154	144	154	123	123
154	128	128	154	125	120	154	125
180	175	225	130	130	120	148	143
215	235	230	235	215	125	215	135
154	154	145	155	215	215	145	225

(a)

214	189	165	122	158	151	134	220
134	215	163	143	143	220	210	225
230	220	228	154	144	154	123	123
154	128	128	154	125	120	154	125
180	175	225	130	130	120	148	143
215	235	230	235	215	125	215	135
154	154	145	155	215	215	145	225

(b)

Figure 7-3. *Grayscale and RGB images*

Convolutional neural networks accept three-dimensional tensors of shape as input: the image height, weight, and depth (Figure 7-4). The depth of the image is the number of channels. For grayscale images the depth is 1, while for RGB images the depth is 3. In Figure 7-4 the input shape of the image is (7, 8, 3).

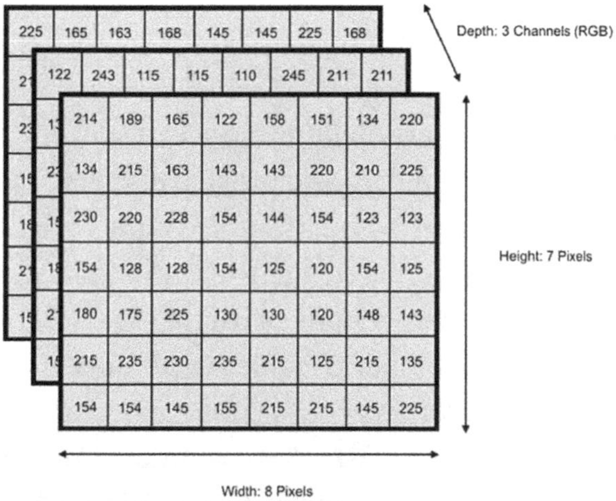

Figure 7-4. *Input shape of an individual image*

Convolutional Neural Network Architecture

Figure 7-5 shows what a typical CNN architecture looks like. Convolutional neural networks consist of several layers, each trying to identify and learn various features. The main types of layers are convolutional layer, pooling layer, and fully connected layer. The layers can be further classified into two main categories: feature detection layers and classification layers.

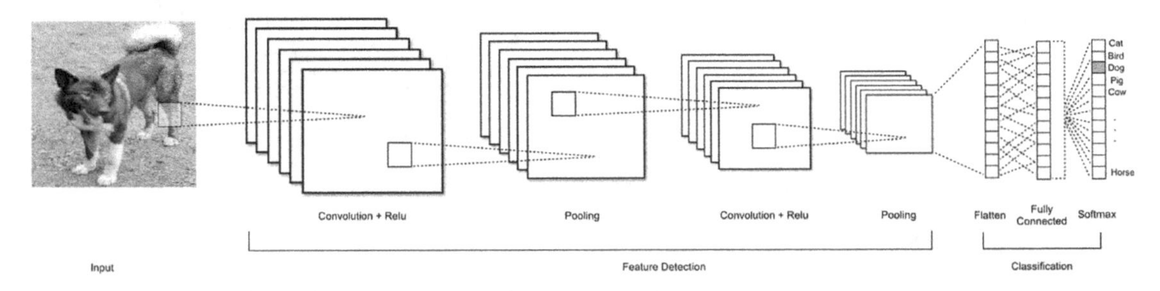

Figure 7-5. *A CNN architecture for classifying animal images*[xxiii]

Feature Detection Layers

Convolutional Layer

The convolutional layer is the primary building block of a convolutional neural network. The convolutional layer runs the input images through a series of convolutional kernels or filter that activates certain features from the input images. A *convolution* is a mathematical operation that performs element-wise multiplication at each location where the input element and the kernel element overlap as the kernel slides (or strides) across the input feature map. The results are added to produce the output in the present location. The process is repeated for every stride, generating the final result called the output feature map.[xxiv] Figure 7-6 shows a 3 x 3 convolutional kernel. Figure 7-7 shows a 2D convolution where our 3 x 3 kernel is applied to a 5 x 5 input feature map, resulting in a 3 x 3 output feature map. We use a stride of 1, which means that the kernel moves one pixel from its current position to the next.

2	0	1
1	2	0
2	2	1

Figure 7-6. *A 3 x 3 convolutional kernel*

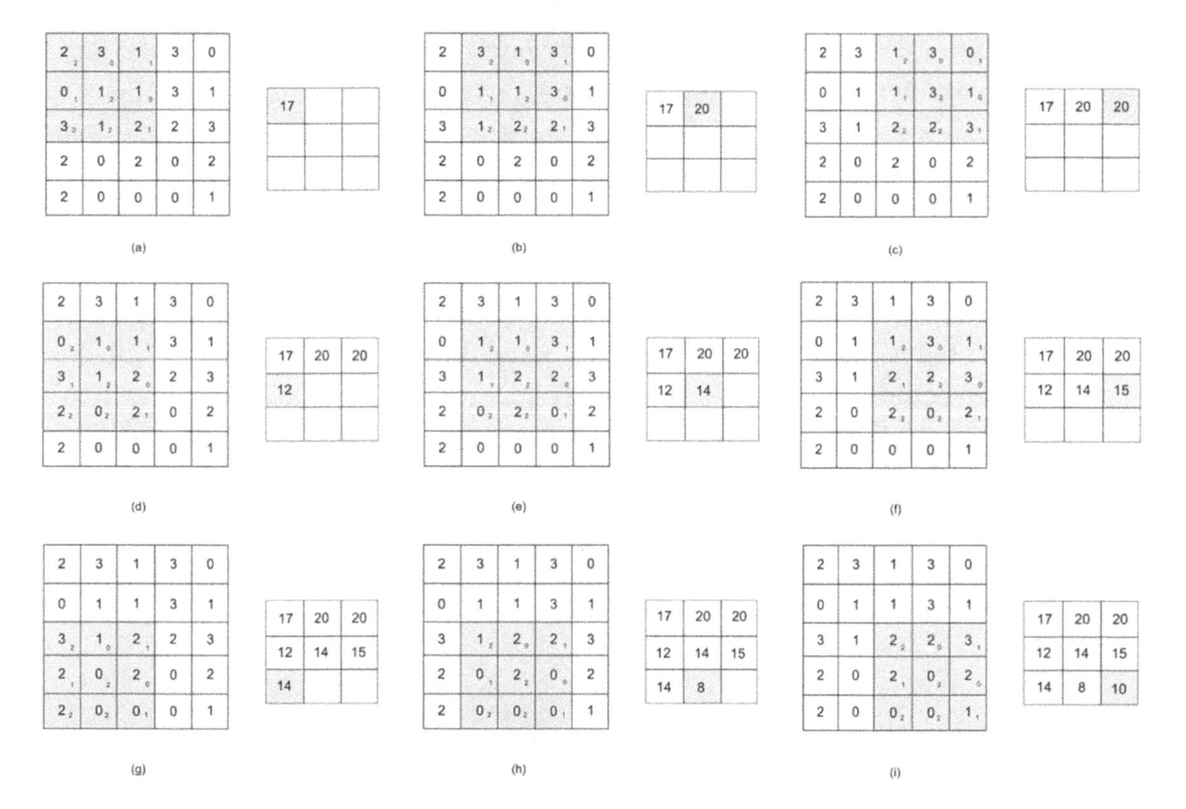

Figure 7-7. *A 2D convolution where a 3 x 3 kernel is applied to a 5 x 5 input feature map (25 features), resulting in a 3 x 3 output feature map with 9 output features*

For an RGB image, a 3D convolution involves a kernel for each input channel (3 kernels in total). The results are all added to form one output. The bias term, used to help the activation function better fit the data, is added to create the final output feature map.

Rectified Linear Unit (ReLU) Activation Function

It is common practice to include an activation layer after each convolutional layer. The activation function can also be specified through the activation argument supported by all forward layers. Activation functions convert an input of a node to an output signal that is used as input to the next layer. Activation functions make neural networks more powerful by allowing it to learn more complex data such as images, audio, and text. It does this by introducing nonlinear properties to our neural network. Without activation function, neural networks would simply be an overly complicated linear regression model that can only handle simple, linear problems.[xxv] Sigmoid and tanh are also common activation layers. Rectified linear unit (ReLU) is the most popular and preferred activation function for most CNNs. ReLU returns the value for any positive input it receives but returns 0 if it receives a negative value. ReLU has been found to help accelerate model training while not significantly affecting accuracy.

Pooling Layer

Pooling layers reduce computational complexity and the parameter count by shrinking the dimension of the input image. By reducing dimensionality, pooling layers also help control overfitting. It is also common to insert a pooling layer after every convolution layer. There are two main kinds of pooling: average pooling and max pooling. Average pooling uses the average of all values from each pooling region (Figure 7-8(b)), while max pooling uses the maximum value (Figure 7-8(a)). In some architectures using larger strides is preferred than using pooling layers.

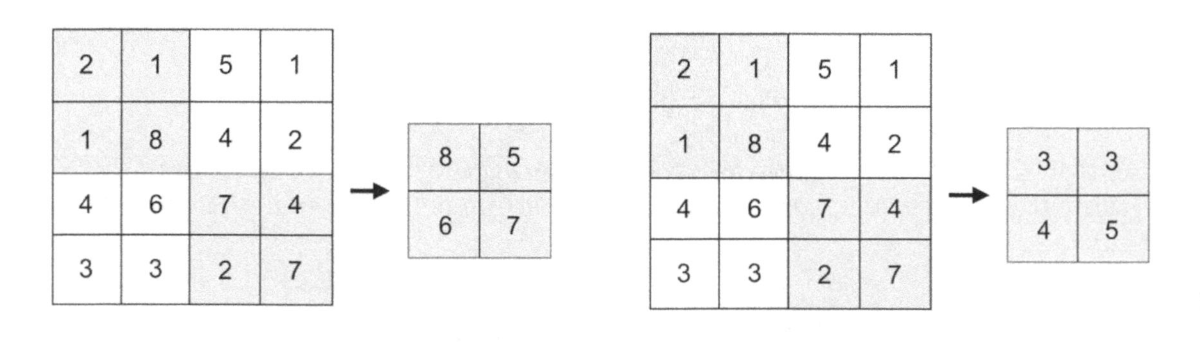

Figure 7-8. *Max and average pooling*

Classification Layers

Flatten Layer

The flatten layer converts a two-dimensional matrix into a one-dimensional vector before the data is fed into the fully connected dense layer.

Fully Connected (Dense) Layer

The fully connected layer, also known as the dense layer, receives the data from the flatten layer and outputs a vector containing the probabilities for each class.

Dropout Layer

A dropout layer randomly deactivates some of the neurons in the network during training. Dropout is a form of regularization that helps reduce model complexity and prevents overfitting.

Softmax and Sigmoid Functions

The final dense layer provides the classification output. It uses either a softmax function for multiclass classification tasks or a sigmoid function for binary classification tasks.

Deep Learning Frameworks

This section provides a quick overview of some of the most popular deep learning frameworks currently available. This is by no means an exhaustive list. Most of these frameworks have more or less the same features. They differ in the level of community and industry adoption, popularity, and size of ecosystem.

TensorFlow

TensorFlow is currently the most popular deep learning framework. It was developed at Google as a replacement for Theano. Some of the original developers of Theano went to Google and developed TensorFlow. TensorFlow provides a Python API that runs on top of an engine developed in C/C++.

Theano

Theano was one of the first open source deep learning frameworks. It was originally released in 2007 by Montreal Institute for Learning Algorithms (MILA) at the University of Montreal. In September 2017 Yoshua Bengio officially announced that development on Theano would end.[xxvi]

PyTorch

PyTorch is an open source deep learning library developed by Facebook Artificial Intelligence Research (FAIR) Group. PyTorch user adoption has been recently surging and is the third most popular framework, behind TensorFlow and Keras.

DeepLearning4J

DeepLearning4J is a deep learning library written in Java. It is compatible with JVM-based languages such as Scala and integrates with Spark and Hadoop. It uses ND4J, its own open source scientific computing library instead of Breeze. DeepLearning4J is a good option for developers who prefer Java or Scala over Python for deep learning (although DeepLearning4J has a Python API that uses Keras).

CNTK

CNTK, also known as Microsoft Cognitive Toolkit, is a deep learning library developed by Microsoft Research and open sourced in April 2015. It uses a sequence of computational steps using a directed graph to describe a neural network. CNTK supports distributed deep learning across several GPUs and servers.

Keras

Keras is a high-level deep learning framework developed at Google by Francois Chollet. It offers a simple and modular API that can run on top of TensorFlow, Theano, and CNTK. It enjoys widespread industry and community adoption and a vibrant ecosystem. Keras models can be deployed on iOS and Android devices, in the browser via Keras.js and WebDNN, on Google Cloud, on the JVM through a DL4J import feature from

Skymind, and even on Raspberry Pi. Keras development is mainly backed by Google but also counts Microsoft, Amazon, Nvidia, Uber, and Apple as key project contributors. Like other frameworks, it has excellent multi-GPU support. For the most complex deep neural networks, Keras supports distributed training through Horovod, GPU clusters on Google Cloud, and Spark with Dist-Keras and Elephas.

Deep Learning with Keras

We will use the Keras Python API with TensorFlow as the back end for all of our deep learning examples. We will use PySpark with Elephas and Distributed Keras for our distributed deep learning examples later in the chapter.

Multiclass Classification Using the Iris Dataset

We'll use a familiar dataset for our first neural network classification[xxvii] task (Listing 7-1). As discussed earlier, we will use Python for all of our examples in this chapter.

Listing 7-1. Multiclass Classification with Keras Using the Iris Dataset

```
import numpy as np

from keras.models import Sequential
from keras.optimizers import Adam
from keras.layers import Dense

from sklearn.model_selection import train_test_split
from sklearn.preprocessing import OneHotEncoder
from sklearn.datasets import load_iris

# Load the Iris dataset.
iris_data = load_iris()

x = iris_data.data

# Inspect the data. We'll truncate
# the output for brevity.
```

```
print(x)
[[5.1 3.5 1.4 0.2]
 [4.9 3.  1.4 0.2]
 [4.7 3.2 1.3 0.2]
 [4.6 3.1 1.5 0.2]
 [5.  3.6 1.4 0.2]
 [5.4 3.9 1.7 0.4]
 [4.6 3.4 1.4 0.3]
 [5.  3.4 1.5 0.2]
 [4.4 2.9 1.4 0.2]
 [4.9 3.1 1.5 0.1]]

# Convert the class label into a single column.
y = iris_data.target.reshape(-1, 1)

# Inspect the class labels. We'll truncate
# the output for brevity.

print(y)
[[0]
 [0]
 [0]
 [0]
 [0]
 [0]
 [0]
 [0]
 [0]
 [0]]

# It's considered best practice to one-hot encode the class labels
# when using neural networks for multiclass classification tasks.

encoder = OneHotEncoder(sparse=False)
enc_y = encoder.fit_transform(y)
print(enc_y)
```

```
[[1. 0. 0.]
 [1. 0. 0.]
 [1. 0. 0.]
 [1. 0. 0.]
 [1. 0. 0.]
 [1. 0. 0.]
 [1. 0. 0.]
 [1. 0. 0.]
 [1. 0. 0.]
 [1. 0. 0.]]

# Split the data for training and testing.
# 70% for training dataset, 30% for test dataset.

train_x, test_x, train_y, test_y = train_test_split(x, enc_y, test_size=0.30)

# Define the model.

# We instantiate a Sequential model consisting of a linear stack of layers.
# Since we are dealing with a one-dimensional feature vector, we will use
# dense layers in our network. We use convolutional layers later in the
# chapter when working with images.

# We use ReLU as the activation function on our fully connected layers.
# We can name any layer by passing the name argument. We pass the number
# of features to the input_shape argument, which is 4 (petal_length,
# petal_width, sepal_length, sepal_width)

model = Sequential()
model.add(Dense(10, input_shape=(4,), activation='relu', name='fclayer1'))
model.add(Dense(10, activation='relu', name='fclayer2'))

# We use softmax activation function in our output layer. As discussed,
# using softmax activation layer allows the model to perform multiclass
# classification. For binary classification, use the sigmoid activation
# function instead. We specify the number of class, which in our case is 3.
```

```
model.add(Dense(3, activation='softmax', name='output'))

# Compile the model. We use categorical_crossentropy for multiclass
# classification. For binary classification, use binary_crossentropy.

model.compile(loss='categorical_crossentropy', optimizer='adam',
metrics=['accuracy'])

# Display the model summary.

print(model.summary())

Model: "sequential_1"
_________________________________________________________________
Layer (type)                 Output Shape              Param #
=================================================================
fclayer1 (Dense)             (None, 10)                50
_________________________________________________________________
fclayer2 (Dense)             (None, 10)                110
_________________________________________________________________
output (Dense)               (None, 3)                 33
=================================================================
Total params: 193
Trainable params: 193
Non-trainable params: 0
_________________________________________________________________
None

# Train the model on training dataset with epochs set to 100
# and batch size to 5. The output is edited for brevity.

model.fit(train_x, train_y, verbose=2, batch_size=5, epochs=100)
Epoch 1/100
 - 2s - loss: 1.3349 - acc: 0.3333
Epoch 2/100
 - 0s - loss: 1.1220 - acc: 0.2952
Epoch 3/100
 - 0s - loss: 1.0706 - acc: 0.3429
```

```
Epoch 4/100
 - 0s - loss: 1.0511 - acc: 0.3810
Epoch 5/100
 - 0s - loss: 1.0353 - acc: 0.3810
Epoch 6/100
 - 0s - loss: 1.0175 - acc: 0.3810
Epoch 7/100
 - 0s - loss: 1.0013 - acc: 0.4000
Epoch 8/100
 - 0s - loss: 0.9807 - acc: 0.4857
Epoch 9/100
 - 0s - loss: 0.9614 - acc: 0.6667
Epoch 10/100
 - 0s - loss: 0.9322 - acc: 0.6857
...
Epoch 97/100
 - 0s - loss: 0.1510 - acc: 0.9524
Epoch 98/100
 - 0s - loss: 0.1461 - acc: 0.9810
Epoch 99/100
 - 0s - loss: 0.1423 - acc: 0.9810
Epoch 100/100
 - 0s - loss: 0.1447 - acc: 0.9810
<keras.callbacks.History object at 0x7fbb93a50510>

# Test on test dataset.
results = model.evaluate(test_x, test_y)
45/45 [==============================] - 0s 586us/step

print('Accuracy: {:4f}'.format(results[1]))
Accuracy: 0.933333
```

That was fun. However, neural networks really shine when used in more complicated problems that involve unstructured data such as images. For our next example, we will use convolutional neural networks to perform handwritten digit recognition.

Handwritten Digit Recognition with MNIST

The MNIST dataset is a database of images of handwritten digits from the US National Institute of Standards and Technology (NIST). The dataset contains 70,000 images: 60,000 images for training and 10,000 images for testing. They are all 28 x 28 grayscale images from 0 to 9 (Figure 7-9). See Listing 7-2.

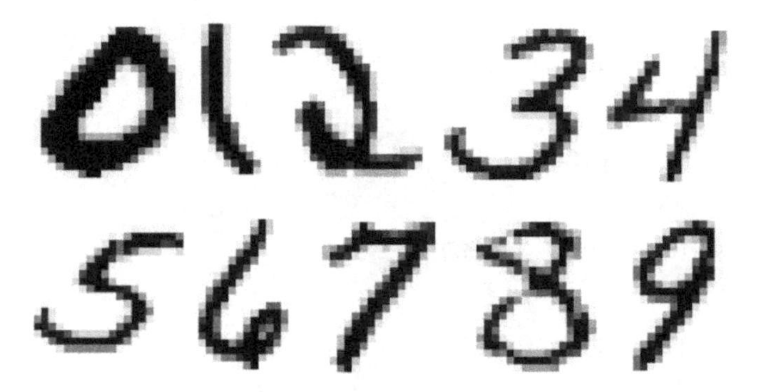

Figure 7-9. *Sample images from the MNIST database of handwritten digits*

Listing 7-2. Handwritten Digit Recognition with MNIST Using Keras

```
import keras
import matplotlib.pyplot as plt

from keras.datasets import mnist
from keras.models import Sequential
from keras.layers import Dense, Conv2D, Dropout, Flatten, MaxPooling2D

# Download MNIST data.

(x_train, y_train), (x_test, y_test) = mnist.load_data()

image_idx = 400
print(y_train[image_idx])

# Inspect the data.

2
plt.imshow(x_train[image_idx], cmap='Greys')
plt.show()
```

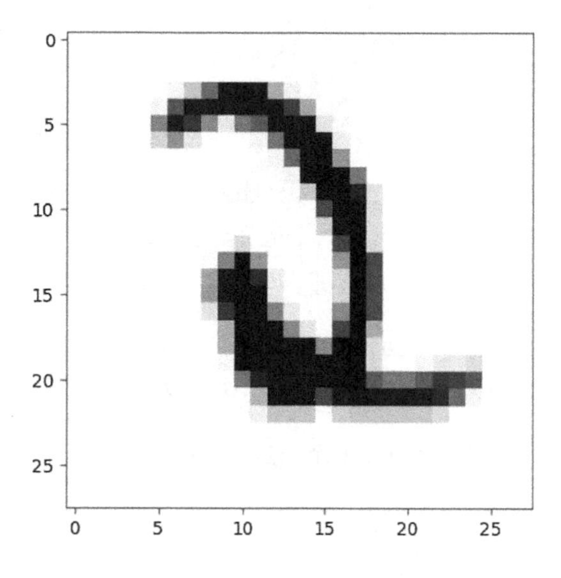


```
# Check another number.

image_idx = 600
print(y_train[image_idx])
9

plt.imshow(x_train[image_idx], cmap='Greys')
plt.show()
```

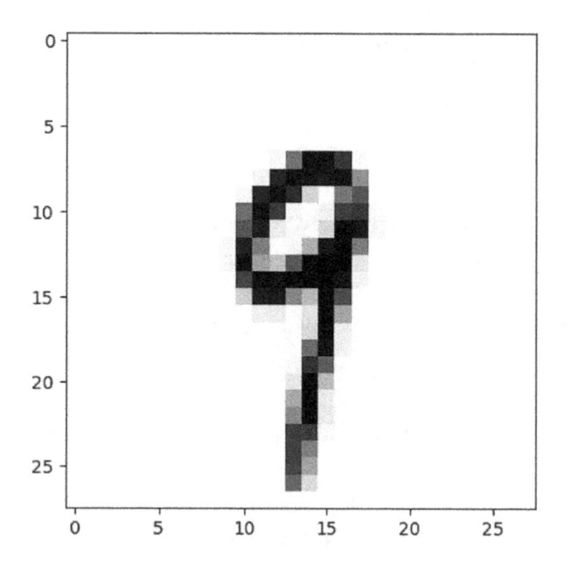

```
# Get the "shape" of the dataset.
x_train.shape
(60000, 28, 28)

# Let's reshape the array to 4-dims so that it can work with Keras.
# The images are 28 x 28 grayscale. The last parameter, 1, indicates
# that images are grayscale. For RGB, set the parameter to 3.

x_train = x_train.reshape(x_train.shape[0], 28, 28, 1)
x_test = x_test.reshape(x_test.shape[0], 28, 28, 1)

# Convert the values to float before division.

x_train = x_train.astype('float32')
x_test = x_test.astype('float32')

# Normalize the RGB codes.

x_train /= 255
x_test /= 255
print'x_train shape: ', x_train.shape
x_train shape:  60000, 28, 28, 1

print'No. of images in training dataset: ', x_train.shape[0]
No. of images in training dataset:  60000

print'No. of images in test dataset: ', x_test.shape[0]
No. of images in test dataset:  10000

# Convert class vectors to binary class matrices. We also pass
# the number of classes (10).

y_train = keras.utils.to_categorical(y_train, 10)
y_test = keras.utils.to_categorical(y_test, 10)

# Build the CNN. Create a Sequential model and add the layers.

model = Sequential()
model.add(Conv2D(28, kernel_size=(3,3), input_shape= (28,28,1),
name='convlayer1'))

# Reduce the dimensionality using max pooling.
```

```
model.add(MaxPooling2D(pool_size=(2, 2)))

# Next we need to flatten the two-dimensional arrays into a one-dimensional
feature vector. This will allow us to perform classification.

model.add(Flatten())
model.add(Dense(128, activation='relu',name='fclayer1'))

# Before we classify the data, we use a dropout layer to randomly
# deactivate some of the neurons. Dropout is a form of regularization
# used to help reduce model complexity and prevent overfitting.

model.add(Dropout(0.2))

# We add the final layer. The softmax layer (multinomial logistic
# regression) should have the total number of classes as parameter.
# In this case the number of parameters is the number of digits (0-9)
# which is 10.

model.add(Dense(10,activation='softmax', name='output'))

# Compile the model.

model.compile(optimizer='adam', loss='categorical_crossentropy',
metrics=['accuracy'])

# Train the model.

model.fit(x_train,y_train,batch_size=128, verbose=1, epochs=20)
Epoch 1/20
60000/60000 [==============================] - 11s 180us/step - loss:
0.2742 - acc: 0.9195
Epoch 2/20
60000/60000 [==============================] - 9s 144us/step - loss:
0.1060 - acc: 0.9682
Epoch 3/20
60000/60000 [==============================] - 9s 143us/step - loss:
0.0731 - acc: 0.9781
Epoch 4/20
```

```
60000/60000 [==============================] - 9s 144us/step - loss:
0.0541 - acc: 0.9830
Epoch 5/20
60000/60000 [==============================] - 9s 143us/step - loss:
0.0409 - acc: 0.9877
Epoch 6/20
60000/60000 [==============================] - 9s 144us/step - loss:
0.0337 - acc: 0.9894
Epoch 7/20
60000/60000 [==============================] - 9s 143us/step - loss:
0.0279 - acc: 0.9910
Epoch 8/20
60000/60000 [==============================] - 9s 144us/step - loss:
0.0236 - acc: 0.9922
Epoch 9/20
60000/60000 [==============================] - 9s 143us/step - loss:
0.0200 - acc: 0.9935
Epoch 10/20
60000/60000 [==============================] - 9s 144us/step - loss:
0.0173 - acc: 0.9940
Epoch 11/20
60000/60000 [==============================] - 9s 143us/step - loss:
0.0163 - acc: 0.9945
Epoch 12/20
60000/60000 [==============================] - 9s 143us/step - loss:
0.0125 - acc: 0.9961
Epoch 13/20
60000/60000 [==============================] - 9s 143us/step - loss:
0.0129 - acc: 0.9956
Epoch 14/20
60000/60000 [==============================] - 9s 144us/step - loss:
0.0125 - acc: 0.9958
Epoch 15/20
```

```
60000/60000 [==============================] - 9s 144us/step - loss:
0.0102 - acc: 0.9968
Epoch 16/20
60000/60000 [==============================] - 9s 143us/step - loss:
0.0101 - acc: 0.9964
Epoch 17/20
60000/60000 [==============================] - 9s 143us/step - loss:
0.0096 - acc: 0.9969
Epoch 18/20
60000/60000 [==============================] - 9s 143us/step - loss:
0.0096 - acc: 0.9968
Epoch 19/20
60000/60000 [==============================] - 9s 142us/step - loss:
0.0090 - acc: 0.9972
Epoch 20/20
60000/60000 [==============================] - 9s 144us/step - loss:
0.0097 - acc: 0.9966
<keras.callbacks.History object at 0x7fc63d629850>

# Evaluate the model.

evalscore = model.evaluate(x_test, y_test, verbose=0)
print'Test accuracy: ', evalscore[1]
Test accuracy:  0.9851
print'Test loss: ', evalscore[0]
Test loss:  0.06053220131823819

# Let's try recognizing some digits.

image_idx = 6700
plt.imshow(x_test[image_idx].reshape(28, 28),cmap='Greys')
plt.show()
```

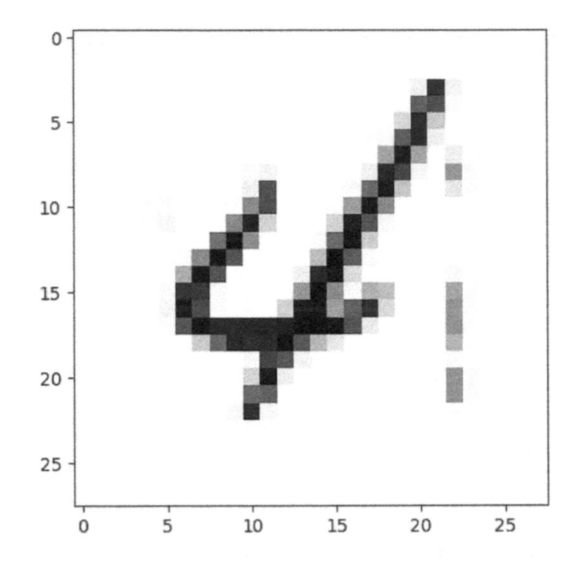

```
pred = model.predict(x_test[image_idx].reshape(1, 28, 28, 1))
print(pred.argmax())
4
```

```
image_idx = 8200
plt.imshow(x_test[image_idx].reshape(28, 28),cmap='Greys')
plt.show()
```

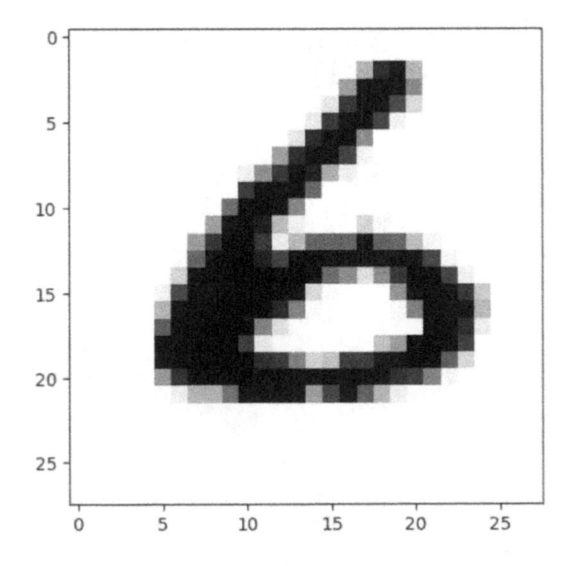

```
pred = model.predict(x_test[image_idx].reshape(1, 28, 28, 1))
print(pred.argmax())
6

image_idx = 8735
plt.imshow(x_test[image_idx].reshape(28, 28),cmap='Greys')
plt.show()
```

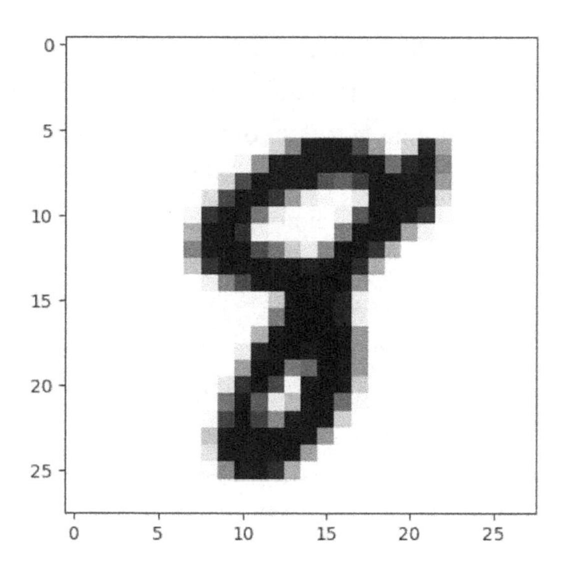


```
pred = model.predict(x_test[image_idx].reshape(1, 28, 28, 1))
print(pred.argmax())
8
```

Congratulations! Our model was able to accurately recognize the digits.

Distributed Deep Learning with Spark

Training complex models like multi-object detectors can take hours, days, or even weeks. In most cases, a single multi-GPU machine is enough to train large models in a reasonable amount of time. For more demanding workloads, spreading computation across multiple machines can dramatically reduce training time, enabling rapid iterative experimentation and accelerating deep learning deployments.

Spark's parallel computing and big data capabilities make it the ideal platform for distributed deep learning. Using Spark for distributed deep learning has additional benefits particularly if you already have an existing Spark cluster. It's convenient to analyze large amount of data stored on the same cluster where the data are stored such as HDFS, Hive, Impala, or HBase. You might also want to share results with other type of workloads running in the same cluster such as business intelligence, machine learning, ETL, and feature engineering.[xxviii]

Model Parallelism vs. Data Parallelism

There are two main approaches to distributed training of neural networks: model parallelism and data parallelism. In data parallelism, each server in a distributed environment gets a complete replica of the model but only a part of the data. Training is performed locally in each server by the replica of the model on the slice of the full dataset (Figure 7-10(a)). In model parallelism, the model is split across different servers (Figure 7-10(b)). Each server is allocated and is responsible for processing a different part of a single neural network, such as a layer.[xxx] Data parallelism is generally more popular due to its simplicity and ease of implementation. However, model parallelism is preferred for training models that are too big to fit in a single machine. DistBelief, Google's framework for large-scale distributed deep learning, supports both model and data parallelism. Horovod, a distributed training framework from Uber, also supports both model and data parallelism.

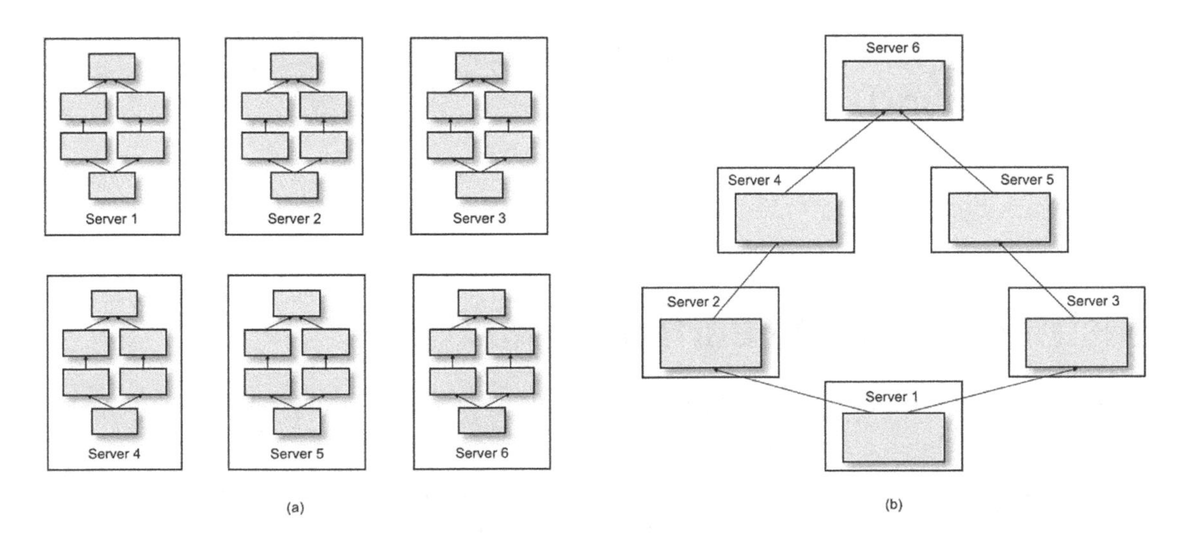

Figure 7-10. *Data parallelism vs. model parallelism*[xxix]

Distributed Deep Learning Frameworks for Spark

Thanks to third-party contributors, even though Spark's deep learning support is still under development, there are several external distributed deep learning frameworks that run on top of Spark. We'll describe the most popular ones.

Deep Learning Pipelines

Deep Learning Pipelines is a third-party package from Databricks (the company founded by the same people who created Spark) that provides deep learning functionality that integrates into the Spark ML Pipelines API. The Deep Learning Pipelines API uses TensorFlow and Keras with TensorFlow as a back end. It includes an ImageSchema that can be used to load images into a Spark DataFrame. It supports transfer learning, distributed hyperparameter tuning, and deploying models as SQL functions.[xxxi] Deep Learning Pipelines is still under active development at the time of this writing.

BigDL

BigDL is a distributed deep learning library for Apache Spark from Intel. It is different from most deep learning frameworks in that it only supports CPU. It uses multithreading and Intel's Math Kernel Library for Deep Neural Networks (Intel MKL-DNN), an open source library for accelerating the performance of deep learning frameworks on Intel architecture. The performance is said to be comparable with conventional GPUs.

CaffeOnSpark

CaffeOnSpark is a deep learning framework developed at Yahoo. It is a distributed extension of Caffe designed to run on top of Spark clusters. CaffeOnSpark is extensively used within Yahoo for content classification and image search.

TensorFlowOnSpark

TensorFlowOnSpark is another deep learning framework developed at Yahoo. It supports distributed TensorFlow inferencing and training using Spark. It integrates with Spark ML Pipelines and supports model and data parallelism and asynchronous and synchronous training.

TensorFrames

TensorFrames is an experimental library that allows TensorFlow to easily work with Spark DataFrames. It supports Scala and Python and provides an efficient way to pass data from Spark to TensorFlow and vice versa.

Elephas

Elephas is a Python library that extends Keras to enable highly scalable distributed deep learning with Spark. Developed by Max Pumperla, Elephas implements distributed deep learning using data parallelism and is known for its ease of use and simplicity. It also supports distributed hyperparameter optimization and distributed training of ensemble models.

Distributed Keras

Distributed Keras (Dist-Keras) is another distributed deep learning framework that run on top of Keras and Spark. It was developed by Joeri Hermans at CERN. It supports several distributed optimization algorithms such as ADAG, Dynamic SGD, Asynchronous Elastic Averaging SGD (AEASGD), Asynchronous Elastic Averaging Momentum SGD (AEAMSGD), and Downpour SGD.

As mentioned earlier, this chapter will focus on Elephas and Distributed Keras.

Elephas: Distributed Deep Learning with Keras and Spark

Elephas is a Python library that extends Keras to enable highly scalable distributed deep learning with Spark. Keras users that have outgrown single multi-GPU machines are looking for ways to scale model training without having to rewrite their existing Keras programs. Elephas (and Distributed Keras) provides an easy way to accomplish this.

Elephas uses data parallelism to distribute training of Keras models across multiple servers. With Elephas, Keras models, data, and parameters are serialized and copied to worker nodes after initialization in the driver. The Spark workers train on their portion of data and send their gradients back and update the master model either synchronously or asynchronously using an optimizer.

The main abstraction of Elephas is the SparkModel. Elephas passes a compiled Keras model to initialize a SparkModel. You can then call the fit method by passing the RDD as training data and options such as the number of epochs, batch sizes, validation splits, and level of verbosity, similar to how you would use Keras.[xxxii] You execute the Python script using spark-submit or pyspark.

```
from elephas.spark_model import SparkModel
from elephas.utils.rdd_utils import to_simple_rdd

rdd = to_simple_rdd(sc, x_train, y_train)

spark_model = SparkModel(model, frequency='epoch', mode='asynchronous')
spark_model.fit(rdd, epochs=10, batch_size=16, verbose=2, validation_
split=0.2)
```

Handwritten Digit Recognition with MNIST Using Elephas with Keras and Spark

We'll use the MNIST dataset for our first example with Elephas.[xxxiii] The code is similar to our previous example with Keras, except that the training data is converted into a Spark RDD and the model is trained using Spark. This way, you can see how easy it is to use Elephas for distributed deep learning. We will use pyspark for our example. See Listing 7-3.

Listing 7-3. Distributed Deep Learning with Elephas, Keras, and Spark

```
# Download MNIST data.

import keras
import matplotlib.pyplot as plt

from keras.datasets import mnist
from keras.models import Sequential
from keras.layers import Dense, Conv2D, Dropout, Flatten, MaxPooling2D
```

```
from elephas.spark_model import SparkModel
from elephas.utils.rdd_utils import to_simple_rdd

(x_train, y_train), (x_test, y_test) = mnist.load_data()

x_train.shape
(60000, 28, 28)

x_train = x_train.reshape(x_train.shape[0], 28, 28, 1)
x_test = x_test.reshape(x_test.shape[0], 28, 28, 1)

x_train = x_train.astype('float32')
x_test = x_test.astype('float32')

x_train /= 255
x_test /= 255

y_train = keras.utils.to_categorical(y_train, 10)
y_test = keras.utils.to_categorical(y_test, 10)

model = Sequential()
model.add(Conv2D(28, kernel_size=(3,3), input_shape= (28,28,1),
name='convlayer1'))
model.add(MaxPooling2D(pool_size=(2, 2)))
model.add(Flatten())
model.add(Dense(128, activation='relu',name='fclayer1'))
model.add(Dropout(0.2))
model.add(Dense(10,activation='softmax', name='output'))

print(model.summary())
Model: "sequential_1"
_________________________________________________________________
Layer (type)                 Output Shape              Param #
=================================================================
convlayer1 (Conv2D)          (None, 26, 26, 28)        280
_________________________________________________________________
max_pooling2d_1 (MaxPooling2 (None, 13, 13, 28)        0
_________________________________________________________________
```

```
flatten_1 (Flatten)          (None, 4732)          0
_________________________________________________________________
fclayer1 (Dense)             (None, 128)           605824
_________________________________________________________________
dropout_1 (Dropout)          (None, 128)           0
_________________________________________________________________
output (Dense)               (None, 10)            1290
=================================================================
Total params: 607,394
Trainable params: 607,394
Non-trainable params: 0
_________________________________________________________________
None

# Compile the model.

model.compile(optimizer='adam', loss='categorical_crossentropy',
metrics=['accuracy'])

# Build RDD from features and labels.
rdd = to_simple_rdd(sc, x_train, y_train)

# Initialize SparkModel from Keras model and Spark context.
spark_model = SparkModel(model, frequency='epoch', mode='asynchronous')

# Train the Spark model.
spark_model.fit(rdd, epochs=20, batch_size=128, verbose=1, validation_
split=0.2)

# The output is edited for brevity.
15104/16051 [===========================>..] - ETA: 0s - loss: 0.0524 -
acc: 0.9852
 9088/16384 [===============>..............] - ETA: 2s - loss: 0.0687 -
acc: 0.9770
 9344/16384 [================>.............] - ETA: 2s - loss: 0.0675 -
acc: 0.9774
```

```
15360/16051 [===========================>..] - ETA: 0s - loss: 0.0520 - 
acc: 0.9852
 9600/16384 [================>.............] - ETA: 2s - loss: 0.0662 - 
acc: 0.9779
15616/16051 [============================>.] - ETA: 0s - loss: 0.0516 - 
acc: 0.9852
 9856/16384 [=================>............] - ETA: 1s - loss: 0.0655 - 
acc: 0.9781
15872/16051 [============================>.] - ETA: 0s - loss: 0.0510 - 
acc: 0.9854
10112/16384 [=================>............] - ETA: 1s - loss: 0.0646 - 
acc: 0.9782
10368/16384 [=================>............] - ETA: 1s - loss: 0.0642 - 
acc: 0.9784
10624/16384 [==================>...........] - ETA: 1s - loss: 0.0645 - 
acc: 0.9784
10880/16384 [==================>...........] - ETA: 1s - loss: 0.0643 - 
acc: 0.9787
11136/16384 [===================>..........] - ETA: 1s - loss: 0.0633 - 
acc: 0.9790
11392/16384 [===================>..........] - ETA: 1s - loss: 0.0620 - 
acc: 0.9795
16051/16051 [==============================] - 6s 370us/step - loss: 
0.0509 - acc: 0.9854 - val_loss: 0.0593 - val_acc: 0.9833
127.0.0.1 - - [01/Sep/2019 23:18:57] "POST /update HTTP/1.1" 200 -

11648/16384 [====================>.........] - ETA: 1s - loss: 0.0623 - 
acc: 0.9794
[Stage 0:==========================================>                (2 + 1)
/ 3]794
12288/16384 [=====================>........] - ETA: 1s - loss: 0.0619 - 
acc: 0.9798
12672/16384 [======================>.......] - ETA: 1s - loss: 0.0615 - 
acc: 0.9799
13056/16384 [======================>.......] - ETA: 0s - loss: 0.0610 - 
acc: 0.9799
```

```
13440/16384 [=======================>......] - ETA: 0s - loss: 0.0598 -
acc: 0.9803
13824/16384 [========================>.....] - ETA: 0s - loss: 0.0588 -
acc: 0.9806
14208/16384 [=========================>....] - ETA: 0s - loss: 0.0581 -
acc: 0.9808
14592/16384 [=========================>....] - ETA: 0s - loss: 0.0577 -
acc: 0.9809
14976/16384 [==========================>...] - ETA: 0s - loss: 0.0565 -
acc: 0.9812
15360/16384 [===========================>..] - ETA: 0s - loss: 0.0566 -
acc: 0.9811
15744/16384 [===========================>..] - ETA: 0s - loss: 0.0564 -
acc: 0.9813
16128/16384 [============================>.] - ETA: 0s - loss: 0.0557 -
acc: 0.9815
16384/16384 [==============================] - 5s 277us/step - loss:
0.0556 - acc: 0.9815 - val_loss: 0.0906 - val_acc: 0.9758
127.0.0.1 - - [01/Sep/2019 23:18:58] "POST /update HTTP/1.1" 200 -
>>> Async training complete.
127.0.0.1 - - [01/Sep/2019 23:18:58] "GET /parameters HTTP/1.1" 200 -

# Evaluate the Spark model.

evalscore = spark_model.master_network.evaluate(x_test, y_test, verbose=2)
print'Test accuracy: ', evalscore[1]
Test accuracy:  0.9644
print'Test loss: ', evalscore[0]
Test loss:  0.12604748902269639

# Perform test digit recognition using our Spark model.

image_idx = 6700
plt.imshow(x_test[image_idx].reshape(28, 28),cmap='Greys')
plt.show()
```

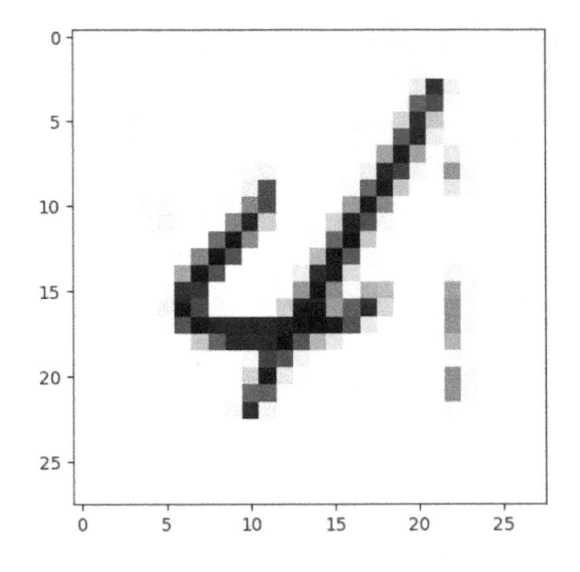


```
pred = spark_model.predict(x_test[image_idx].reshape(1, 28, 28, 1))
print(pred.argmax())
4

image_idx = 8200
plt.imshow(x_test[image_idx].reshape(28, 28),cmap='Greys')
plt.show()
```

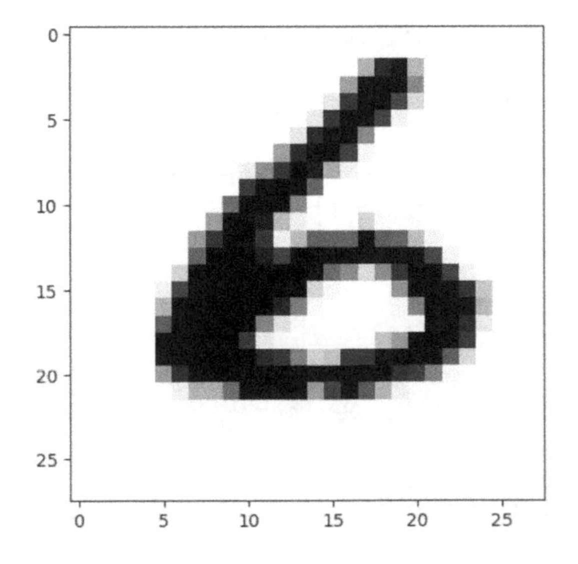

```
pred = spark_model.predict(x_test[image_idx].reshape(1, 28, 28, 1))
print(pred.argmax())
6

image_idx = 8735
plt.imshow(x_test[image_idx].reshape(28, 28),cmap='Greys')
plt.show()
```

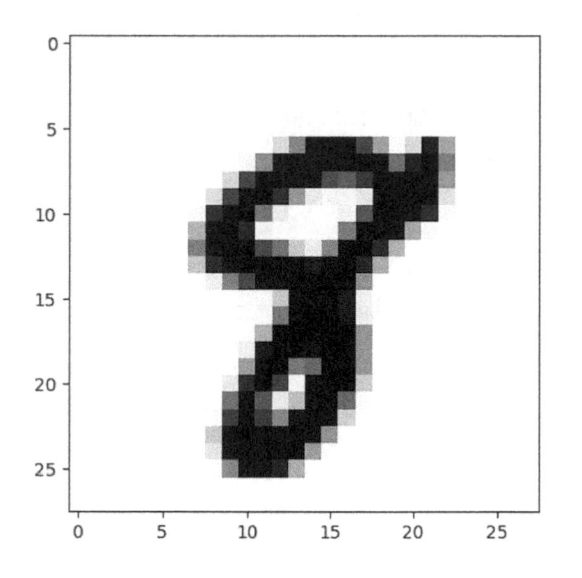


```
pred = spark_model.predict(x_test[image_idx].reshape(1, 28, 28, 1))
print(pred.argmax())
8
```

In our example, we used Python to generate the RDD from numpy arrays. This may not be feasible for truly large datasets. If your data can't fit into memory, it is best to use Spark to create the RDD by reading directly from a distributed storage engine (such as HDFS or S3) and performing all preprocessing using Spark MLlib's transformers and estimators. By leveraging Spark's distributed processing capabilities to generate the RDD, you have a fully distributed deep learning platform.

Elephas also supports model training using Spark MLlib estimators with Spark DataFrames. You can run the estimator as part of a bigger Spark MLlib pipeline. See Listing 7-4.

Listing 7-4. Training a Model with a Spark ML Estimator Using a DataFrame

```
import keras
import matplotlib.pyplot as plt

from keras.datasets import mnist
from keras.models import Sequential
from keras.layers import Dense, Dropout
from keras import optimizers

from pyspark.sql.functions import rand
from pyspark.mllib.evaluation import MulticlassMetrics
from pyspark.ml import Pipeline

from elephas.ml_model import ElephasEstimator
from elephas.ml.adapter import to_data_frame

# Download MNIST data.
(x_train, y_train), (x_test, y_test) = mnist.load_data()

x_train.shape
#(60000, 28, 28)

# We will be using only dense layers for our network. Let's flatten
# the grayscale 28x28 images to a 784 vector (28x28x1 = 784).

x_train = x_train.reshape(60000, 784)
x_test = x_test.reshape(10000, 784)

x_train = x_train.astype('float32')
x_test = x_test.astype('float32')

x_train /= 255
x_test /= 255

y_train = keras.utils.to_categorical(y_train, 10)
y_test = keras.utils.to_categorical(y_test, 10)

# Since Spark DataFrames can't be created from three-dimensional
# data, we need to use dense layers when using Elephas and Keras
# with Spark DataFrames. Use RDDs if you need to use convolutional
```

```
# layers with Elephas.

model = Sequential()
model.add(Dense(128, input_dim=784, activation='relu', name='fclayer1'))
model.add(Dropout(0.2))
model.add(Dense(128, activation='relu',name='fclayer2'))
model.add(Dropout(0.2))
model.add(Dense(10,activation='softmax', name='output'))

# Build Spark DataFrames from features and labels.

df = to_data_frame(sc, x_train, y_train, categorical=True)

df.show(20,50)
+--------------------------------------------------+-----+
|                                          features|label|
+--------------------------------------------------+-----+
|[0.0,0.0,0.0,0.0,0.0,0.0,0.0,0.0,0.0,0.0,0.0,0....|  5.0|
|[0.0,0.0,0.0,0.0,0.0,0.0,0.0,0.0,0.0,0.0,0.0,0....|  0.0|
|[0.0,0.0,0.0,0.0,0.0,0.0,0.0,0.0,0.0,0.0,0.0,0....|  4.0|
|[0.0,0.0,0.0,0.0,0.0,0.0,0.0,0.0,0.0,0.0,0.0,0....|  1.0|
|[0.0,0.0,0.0,0.0,0.0,0.0,0.0,0.0,0.0,0.0,0.0,0....|  9.0|
|[0.0,0.0,0.0,0.0,0.0,0.0,0.0,0.0,0.0,0.0,0.0,0....|  2.0|
|[0.0,0.0,0.0,0.0,0.0,0.0,0.0,0.0,0.0,0.0,0.0,0....|  1.0|
|[0.0,0.0,0.0,0.0,0.0,0.0,0.0,0.0,0.0,0.0,0.0,0....|  3.0|
|[0.0,0.0,0.0,0.0,0.0,0.0,0.0,0.0,0.0,0.0,0.0,0....|  1.0|
|[0.0,0.0,0.0,0.0,0.0,0.0,0.0,0.0,0.0,0.0,0.0,0....|  4.0|
|[0.0,0.0,0.0,0.0,0.0,0.0,0.0,0.0,0.0,0.0,0.0,0....|  3.0|
|[0.0,0.0,0.0,0.0,0.0,0.0,0.0,0.0,0.0,0.0,0.0,0....|  5.0|
|[0.0,0.0,0.0,0.0,0.0,0.0,0.0,0.0,0.0,0.0,0.0,0....|  3.0|
|[0.0,0.0,0.0,0.0,0.0,0.0,0.0,0.0,0.0,0.0,0.0,0....|  6.0|
|[0.0,0.0,0.0,0.0,0.0,0.0,0.0,0.0,0.0,0.0,0.0,0....|  1.0|
|[0.0,0.0,0.0,0.0,0.0,0.0,0.0,0.0,0.0,0.0,0.0,0....|  7.0|
|[0.0,0.0,0.0,0.0,0.0,0.0,0.0,0.0,0.0,0.0,0.0,0....|  2.0|
|[0.0,0.0,0.0,0.0,0.0,0.0,0.0,0.0,0.0,0.0,0.0,0....|  8.0|
|[0.0,0.0,0.0,0.0,0.0,0.0,0.0,0.0,0.0,0.0,0.0,0....|  6.0|
|[0.0,0.0,0.0,0.0,0.0,0.0,0.0,0.0,0.0,0.0,0.0,0....|  9.0|
+--------------------------------------------------+-----+
```

```
only showing top 20 rows

test_df = to_data_frame(sc, x_test, y_test, categorical=True)

test_df.show(20,50)

+--------------------------------------------------+-----+
|                                          features|label|
+--------------------------------------------------+-----+
|[0.0,0.0,0.0,0.0,0.0,0.0,0.0,0.0,0.0,0.0,0.0,0....|  7.0|
|[0.0,0.0,0.0,0.0,0.0,0.0,0.0,0.0,0.0,0.0,0.0,0....|  2.0|
|[0.0,0.0,0.0,0.0,0.0,0.0,0.0,0.0,0.0,0.0,0.0,0....|  1.0|
|[0.0,0.0,0.0,0.0,0.0,0.0,0.0,0.0,0.0,0.0,0.0,0....|  0.0|
|[0.0,0.0,0.0,0.0,0.0,0.0,0.0,0.0,0.0,0.0,0.0,0....|  4.0|
|[0.0,0.0,0.0,0.0,0.0,0.0,0.0,0.0,0.0,0.0,0.0,0....|  1.0|
|[0.0,0.0,0.0,0.0,0.0,0.0,0.0,0.0,0.0,0.0,0.0,0....|  4.0|
|[0.0,0.0,0.0,0.0,0.0,0.0,0.0,0.0,0.0,0.0,0.0,0....|  9.0|
|[0.0,0.0,0.0,0.0,0.0,0.0,0.0,0.0,0.0,0.0,0.0,0....|  5.0|
|[0.0,0.0,0.0,0.0,0.0,0.0,0.0,0.0,0.0,0.0,0.0,0....|  9.0|
|[0.0,0.0,0.0,0.0,0.0,0.0,0.0,0.0,0.0,0.0,0.0,0....|  0.0|
|[0.0,0.0,0.0,0.0,0.0,0.0,0.0,0.0,0.0,0.0,0.0,0....|  6.0|
|[0.0,0.0,0.0,0.0,0.0,0.0,0.0,0.0,0.0,0.0,0.0,0....|  9.0|
|[0.0,0.0,0.0,0.0,0.0,0.0,0.0,0.0,0.0,0.0,0.0,0....|  0.0|
|[0.0,0.0,0.0,0.0,0.0,0.0,0.0,0.0,0.0,0.0,0.0,0....|  1.0|
|[0.0,0.0,0.0,0.0,0.0,0.0,0.0,0.0,0.0,0.0,0.0,0....|  5.0|
|[0.0,0.0,0.0,0.0,0.0,0.0,0.0,0.0,0.0,0.0,0.0,0....|  9.0|
|[0.0,0.0,0.0,0.0,0.0,0.0,0.0,0.0,0.0,0.0,0.0,0....|  7.0|
|[0.0,0.0,0.0,0.0,0.0,0.0,0.0,0.0,0.0,0.0,0.0,0....|  3.0|
|[0.0,0.0,0.0,0.0,0.0,0.0,0.0,0.0,0.0,0.0,0.0,0....|  4.0|
+--------------------------------------------------+-----+
only showing top 20 rows

# Set and serialize optimizer.

sgd = optimizers.SGD(lr=0.01)
optimizer_conf = optimizers.serialize(sgd)

# Initialize Spark ML Estimator.
```

```
estimator = ElephasEstimator()
estimator.set_keras_model_config(model.to_yaml())
estimator.set_optimizer_config(optimizer_conf)
estimator.set_epochs(25)
estimator.set_batch_size(64)
estimator.set_categorical_labels(True)
estimator.set_validation_split(0.10)
estimator.set_nb_classes(10)
estimator.set_mode("synchronous")
estimator.set_loss("categorical_crossentropy")
estimator.set_metrics(['acc'])

# Fit a model.

pipeline = Pipeline(stages=[estimator])
pipeline_model = pipeline.fit(df)

# Evaluate the fitted pipeline model on test data.

prediction = pipeline_model.transform(test_df)
df2 = prediction.select("label", "prediction")
df2.show(20)

+-----+----------+
|label|prediction|
+-----+----------+
|  7.0|       7.0|
|  2.0|       2.0|
|  1.0|       1.0|
|  0.0|       0.0|
|  4.0|       4.0|
|  1.0|       1.0|
|  4.0|       4.0|
|  9.0|       9.0|
|  5.0|       6.0|
|  9.0|       9.0|
|  0.0|       0.0|
|  6.0|       6.0|
```

```
|   9.0|         9.0|
|   0.0|         0.0|
|   1.0|         1.0|
|   5.0|         5.0|
|   9.0|         9.0|
|   7.0|         7.0|
|   3.0|         2.0|
|   4.0|         4.0|
+-----+----------+
only showing top 20 rows

prediction_and_label= df2.rdd.map(lambda row: (row.label, row.prediction))
metrics = MulticlassMetrics(prediction_and_label)
print(metrics.precision())
0.757
```

Distributed Keras (Dist-Keras)

Distributed Keras (Dist-Keras) is another distributed deep learning framework that runs on top of Keras and Spark. It was developed by Joeri Hermans at CERN. It supports several distributed optimization algorithms such as ADAG, Dynamic SGD, Asynchronous Elastic Averaging SGD (AEASGD), Asynchronous Elastic Averaging Momentum SGD (AEAMSGD), and Downpour SGD. Dist-Keras includes its own Spark transformers for various data transformations such as ReshapeTransformer, MinMaxTransformer, OneHotTransformer, DenseTransformer, and LabelIndexTransformer, to mention a few. Similar to Elephas, Dist-Keras implements distributed deep learning using data parallelism.

Handwritten Digit Recognition with MNIST Using Dist-Keras with Keras and Spark

For consistency, we will use the MNIST dataset for our Dist-Keras example.[xxxiv] Before running this example, make sure you put the MNIST dataset on HDFS or S3. See Listing 7-5.

Listing 7-5. Distributed Deep Learning with Dist-Keras, Keras, and Spark

```
from distkeras.evaluators import *
from distkeras.predictors import *
from distkeras.trainers import *
from distkeras.transformers import *
from distkeras.utils import *

from keras.layers.convolutional import *
from keras.layers.core import *
from keras.models import Sequential
from keras.optimizers import *

from pyspark import SparkConf
from pyspark import SparkContext

from pyspark.ml.evaluation import MulticlassClassificationEvaluator
from pyspark.ml.feature import OneHotEncoder
from pyspark.ml.feature import StandardScaler
from pyspark.ml.feature import StringIndexer
from pyspark.ml.feature import VectorAssembler

import pwd
import os
# First, set up the Spark variables. You can modify them to your needs.
application_name = "Distributed Keras MNIST Notebook"
using_spark_2 = False
local = False
path_train = "data/mnist_train.csv"
path_test = "data/mnist_test.csv"
if local:
    # Tell master to use local resources.
    master = "local[*]"
    num_processes = 3
    num_executors = 1
```

```
else:
    # Tell master to use YARN.
    master = "yarn-client"
    num_executors = 20
    num_processes = 1

# This variable is derived from the number of cores and executors and will
be used to assign the number of model trainers.
num_workers = num_executors * num_processes

print("Number of desired executors: " + `num_executors`)
print("Number of desired processes / executor: " + `num_processes`)
print("Total number of workers: " + `num_workers`)

# Use the Databricks CSV reader; this has some nice functionality regarding
invalid values.
os.environ['PYSPARK_SUBMIT_ARGS'] = '--packages com.databricks:spark-
csv_2.10:1.4.0 pyspark-shell'

conf = SparkConf()
conf.set("spark.app.name", application_name)
conf.set("spark.master", master)
conf.set("spark.executor.cores", `num_processes`)
conf.set("spark.executor.instances", `num_executors`)
conf.set("spark.executor.memory", "4g")
conf.set("spark.locality.wait", "0")
conf.set("spark.serializer", "org.apache.spark.serializer.KryoSerializer");
conf.set("spark.local.dir", "/tmp/" + get_os_username() + "/dist-keras");

# Check if the user is running Spark 2.0 +
if using_spark_2:
    sc = SparkSession.builder.config(conf=conf) \
            .appName(application_name) \
            .getOrCreate()
else:
    # Create the Spark context.
    sc = SparkContext(conf=conf)
    # Add the missing imports.
```

```
    from pyspark import SQLContext
    sqlContext = SQLContext(sc)

# Check if we are using Spark 2.0.
if using_spark_2:
    reader = sc
else:
    reader = sqlContext
# Read the training dataset.
raw_dataset_train = reader.read.format('com.databricks.spark.csv') \
                          .options(header='true', inferSchema='true') \
                          .load(path_train)
# Read the testing dataset.
raw_dataset_test = reader.read.format('com.databricks.spark.csv') \
                         .options(header='true', inferSchema='true') \
                         .load(path_test)

# First, we would like to extract the desired features from the raw
# dataset. We do this by constructing a list with all desired columns.
# This is identical for the test set.

features = raw_dataset_train.columns
features.remove('label')

# Next, we use Spark's VectorAssembler to "assemble" (create) a vector of
# all desired features.

vector_assembler = VectorAssembler(inputCols=features,
outputCol="features")

# This transformer will take all columns specified in features and create #
an additional column "features" which will contain all the desired
# features aggregated into a single vector.

dataset_train = vector_assembler.transform(raw_dataset_train)
dataset_test = vector_assembler.transform(raw_dataset_test)

# Define the number of output classes.
```

```
nb_classes = 10
encoder = OneHotTransformer(nb_classes, input_col="label", output_
col="label_encoded")
dataset_train = encoder.transform(dataset_train)
dataset_test = encoder.transform(dataset_test)

# Allocate a MinMaxTransformer from Distributed Keras to normalize
# the features.
# o_min -> original_minimum
# n_min -> new_minimum

transformer = MinMaxTransformer(n_min=0.0, n_max=1.0, \
                                o_min=0.0, o_max=250.0, \
                                input_col="features", \
                                output_col="features_normalized")
# Transform the dataset.
dataset_train = transformer.transform(dataset_train)
dataset_test = transformer.transform(dataset_test)

# Keras expects the vectors to be in a particular shape; we can reshape the
# vectors using Spark.
reshape_transformer = ReshapeTransformer("features_normalized", "matrix",
(28, 28, 1))
dataset_train = reshape_transformer.transform(dataset_train)
dataset_test = reshape_transformer.transform(dataset_test)

# Now, create a Keras model.
# Taken from Keras MNIST example.

# Declare model parameters.
img_rows, img_cols = 28, 28
# Number of convolutional filters to use
nb_filters = 32
# Size of pooling area for max pooling
pool_size = (2, 2)
# Convolution kernel size
kernel_size = (3, 3)
input_shape = (img_rows, img_cols, 1)
```

```
# Construct the model.
convnet = Sequential()
convnet.add(Convolution2D(nb_filters, kernel_size[0], kernel_size[1],
                          border_mode='valid',
                          input_shape=input_shape))
convnet.add(Activation('relu'))
convnet.add(Convolution2D(nb_filters, kernel_size[0], kernel_size[1]))
convnet.add(Activation('relu'))
convnet.add(MaxPooling2D(pool_size=pool_size))
convnet.add(Flatten())
convnet.add(Dense(225))
convnet.add(Activation('relu'))
convnet.add(Dense(nb_classes))
convnet.add(Activation('softmax'))

# Define the optimizer and the loss.
optimizer_convnet = 'adam'
loss_convnet = 'categorical_crossentropy'

# Print the summary.
convnet.summary()

# We can also evaluate the dataset in a distributed manner.
# However, for this we need to specify a procedure on how to do this.
def evaluate_accuracy(model, test_set, features="matrix"):
    evaluator = AccuracyEvaluator(prediction_col="prediction_index",
    label_col="label")
    predictor = ModelPredictor(keras_model=model, features_col=features)
    transformer = LabelIndexTransformer(output_dim=nb_classes)
    test_set = test_set.select(features, "label")
    test_set = predictor.predict(test_set)
    test_set = transformer.transform(test_set)
    score = evaluator.evaluate(test_set)

    return score
```

```
# Select the desired columns; this will reduce network usage.
dataset_train = dataset_train.select("features_normalized",
"matrix","label", "label_encoded")
dataset_test = dataset_test.select("features_normalized", "matrix","label",
"label_encoded")

# Keras expects DenseVectors.

dense_transformer = DenseTransformer(input_col="features_normalized",
output_col="features_normalized_dense")
dataset_train = dense_transformer.transform(dataset_train)
dataset_test = dense_transformer.transform(dataset_test)
dataset_train.repartition(num_workers)
dataset_test.repartition(num_workers)

# Assessing the training and test set.
training_set = dataset_train.repartition(num_workers)
test_set = dataset_test.repartition(num_workers)

# Cache them.
training_set.cache()
test_set.cache()

# Precache the training set on the nodes using a simple count.
print(training_set.count())

# Use the ADAG optimizer. You can also use a SingleWorker for testing
# purposes -> traditional nondistributed gradient descent.

trainer = ADAG(keras_model=convnet, worker_optimizer=optimizer_convnet,
loss=loss_convnet, num_workers=num_workers, batch_size=16, communication_
window=5, num_epoch=5, features_col="matrix", label_col="label_encoded")

trained_model = trainer.train(training_set)

print("Training time: " + str(trainer.get_training_time()))
print("Accuracy: " + str(evaluate_accuracy(trained_model, test_set)))
print("Number of parameter server updates: " + str(trainer.parameter_
server.num_updates))
```

Distributing deep learning workloads across multiple machines is not always a good approach. There is overhead in running jobs on distributed environments. Not to mention the amount of time and effort that goes into setting up and maintaining a distributed Spark environment. High-performance multi-GPU machines and cloud instances already allow you to train fairly large models on single machines with good training speed. You may not need a distributed environment at all. In fact, utilizing the ImageDataGenerator class to load data and the fit_generator function to train models in Keras might be sufficient in most cases. Let's explore this option in our next example.

Dogs and Cats Image Classification

In this example, we will use a convolutional neural network to build a dog and cat image classifier. We'll use a popular dataset that was popularized by Francois Chollet, and provided by Microsoft Research and available from Kaggle.[xxxv] Like the MNIST dataset, this dataset is widely used in research. It contains 12,500 cat images and 12,500 dog images, although we will only use 2000 images for each class (4000 total images) to speed up training. We will use 500 cat images and 500 dog images (total 1000 images) for testing. See Figure 7-11.

Figure 7-11. *Sample dogs and cats images from the dataset*

We will use fit_generator to train our Keras model. We will also utilize the ImageDataGenerator class to load data in batches, allowing us to process large amounts of data. This is particularly useful if you have large datasets that can't fit in memory and you don't have access to a distributed environment. An added benefit of using ImageDataGenerator is its ability to perform random data transformation to augment your data, helping the model generalize better and help prevent overfitting.[xxxvi] This example will show how to use large datasets with Keras without using Spark. See Listing 7-6.

Listing 7-6. Using ImageDataGenerator and Fit_Generator

```
import matplotlib.pyplot as plt
import numpy as np
import cv2

from keras.preprocessing.image import ImageDataGenerator
from keras.preprocessing import image
from keras.models import Sequential
from keras.layers import Conv2D, MaxPooling2D
from keras.layers import Activation, Dropout, Flatten, Dense
from keras import backend as K

# The image dimension is 150x150. RGB = 3.

if K.image_data_format() == 'channels_first':
    input_shape = (3, 150, 150)
else:
    input_shape = (150, 150, 3)

model = Sequential()

model.add(Conv2D(32, (3, 3), input_shape=input_shape, activation='relu'))
model.add(MaxPooling2D(pool_size=(2, 2)))

model.add(Conv2D(32, (3, 3), activation='relu'))
model.add(MaxPooling2D(pool_size=(2, 2)))

model.add(Conv2D(64, (3, 3), activation='relu'))
model.add(MaxPooling2D(pool_size=(2, 2)))

model.add(Flatten())
model.add(Dense(64, activation='relu'))
model.add(Dropout(0.5))
model.add(Dense(1, activation='sigmoid'))
```

```
# Compile the model.

model.compile(loss='binary_crossentropy',
              optimizer='rmsprop',
              metrics=['accuracy'])

print(model.summary())

_________________________________________________________________
Layer (type)                 Output Shape              Param #
=================================================================
conv2d_1 (Conv2D)            (None, 148, 148, 32)      896
_________________________________________________________________
max_pooling2d_1 (MaxPooling2 (None, 74, 74, 32)        0
_________________________________________________________________
conv2d_2 (Conv2D)            (None, 72, 72, 32)        9248
_________________________________________________________________
max_pooling2d_2 (MaxPooling2 (None, 36, 36, 32)        0
_________________________________________________________________
conv2d_3 (Conv2D)            (None, 34, 34, 64)        18496
_________________________________________________________________
max_pooling2d_3 (MaxPooling2 (None, 17, 17, 64)        0
_________________________________________________________________
flatten_1 (Flatten)          (None, 18496)             0
_________________________________________________________________
dense_1 (Dense)              (None, 64)                1183808
_________________________________________________________________
dropout_1 (Dropout)          (None, 64)                0
_________________________________________________________________
dense_2 (Dense)              (None, 1)                 65
=================================================================
Total params: 1,212,513
Trainable params: 1,212,513
Non-trainable params: 0
_________________________________________________________________
None
```

```
# We will use the following augmentation configuration for training.

train_datagen = ImageDataGenerator(
    rescale=1. / 255,
    width_shift_range=0.2,
    height_shift_range=0.2,
    horizontal_flip=True)

# The only augmentation for test data is rescaling.

test_datagen = ImageDataGenerator(rescale=1. / 255)

train_generator = train_datagen.flow_from_directory(
    '/data/train',
    target_size=(150, 150),
    batch_size=16,
    class_mode='binary')

Found 4000 images belonging to 2 classes.

validation_generator = test_datagen.flow_from_directory(
    '/data/test',
    target_size=(150, 150),
    batch_size=16,
    class_mode='binary')

Found 1000 images belonging to 2 classes.

# steps_per_epoch should be set to the total number of training sample,
# while validation_steps is set to the number of test samples. We set
# epoch to 15, steps_per_epoch and validation_steps to 100 to expedite
# model training.

model.fit_generator(
    train_generator,
    steps_per_epoch=100,
    epochs=1=25,
    validation_data=validation_generator,
    validation_steps=100)
```

```
Epoch 1/25
100/100 [==============================] - 45s 451ms/step - loss: 0.6439 -
acc: 0.6244 - val_loss: 0.5266 - val_acc: 0.7418
Epoch 2/25
100/100 [==============================] - 44s 437ms/step - loss: 0.6259 -
acc: 0.6681 - val_loss: 0.5577 - val_acc: 0.7304
Epoch 3/25
100/100 [==============================] - 43s 432ms/step - loss: 0.6326 -
acc: 0.6338 - val_loss: 0.5922 - val_acc: 0.7029
Epoch 4/25
100/100 [==============================] - 43s 434ms/step - loss: 0.6538 -
acc: 0.6300 - val_loss: 0.5642 - val_acc: 0.7052
Epoch 5/25
100/100 [==============================] - 44s 436ms/step - loss: 0.6263 -
acc: 0.6600 - val_loss: 0.6725 - val_acc: 0.6746
Epoch 6/25
100/100 [==============================] - 43s 427ms/step - loss: 0.6229 -
acc: 0.6606 - val_loss: 0.5586 - val_acc: 0.7538
Epoch 7/25
100/100 [==============================] - 43s 426ms/step - loss: 0.6470 -
acc: 0.6562 - val_loss: 0.5878 - val_acc: 0.7077
Epoch 8/25
100/100 [==============================] - 43s 429ms/step - loss: 0.6524 -
acc: 0.6437 - val_loss: 0.6414 - val_acc: 0.6539
Epoch 9/25
100/100 [==============================] - 43s 427ms/step - loss: 0.6131 -
acc: 0.6831 - val_loss: 0.5636 - val_acc: 0.7304
Epoch 10/25
100/100 [==============================] - 43s 429ms/step - loss: 0.6293 -
acc: 0.6538 - val_loss: 0.5857 - val_acc: 0.7186
Epoch 11/25
100/100 [==============================] - 44s 437ms/step - loss: 0.6207 -
acc: 0.6713 - val_loss: 0.5467 - val_acc: 0.7279
```

```
Epoch 12/25
100/100 [==============================] - 43s 430ms/step - loss: 0.6131 -
acc: 0.6587 - val_loss: 0.5279 - val_acc: 0.7348
Epoch 13/25
100/100 [==============================] - 43s 428ms/step - loss: 0.6090 -
acc: 0.6781 - val_loss: 0.6221 - val_acc: 0.7054
Epoch 14/25
100/100 [==============================] - 42s 421ms/step - loss: 0.6273 -
acc: 0.6756 - val_loss: 0.5446 - val_acc: 0.7506
Epoch 15/25
100/100 [==============================] - 44s 442ms/step - loss: 0.6139 -
acc: 0.6775 - val_loss: 0.6073 - val_acc: 0.6954
Epoch 16/25
100/100 [==============================] - 44s 441ms/step - loss: 0.6080 -
acc: 0.6806 - val_loss: 0.5365 - val_acc: 0.7437
Epoch 17/25
100/100 [==============================] - 45s 448ms/step - loss: 0.6225 -
acc: 0.6719 - val_loss: 0.5831 - val_acc: 0.6935
Epoch 18/25
100/100 [==============================] - 43s 428ms/step - loss: 0.6124 -
acc: 0.6769 - val_loss: 0.5457 - val_acc: 0.7361
Epoch 19/25
100/100 [==============================] - 43s 430ms/step - loss: 0.6061 -
acc: 0.6844 - val_loss: 0.5587 - val_acc: 0.7399
Epoch 20/25
100/100 [==============================] - 43s 429ms/step - loss: 0.6209 -
acc: 0.6613 - val_loss: 0.5699 - val_acc: 0.7280
Epoch 21/25
100/100 [==============================] - 43s 428ms/step - loss: 0.6252 -
acc: 0.6650 - val_loss: 0.5550 - val_acc: 0.7247
Epoch 22/25
100/100 [==============================] - 43s 429ms/step - loss: 0.6306 -
acc: 0.6594 - val_loss: 0.5466 - val_acc: 0.7236
Epoch 23/25
100/100 [==============================] - 43s 427ms/step - loss: 0.6086 -
acc: 0.6819 - val_loss: 0.5790 - val_acc: 0.6824
```

```
Epoch 24/25
100/100 [==============================] - 43s 425ms/step - loss: 0.6059 -
acc: 0.7000 - val_loss: 0.5433 - val_acc: 0.7197
Epoch 25/25
100/100 [==============================] - 43s 426ms/step - loss: 0.6261 -
acc: 0.6794 - val_loss: 0.5987 - val_acc: 0.7167
<keras.callbacks.History object at 0x7ff72c7c3890>

# We get a 71% validation accuracy. To increase model accuracy, you can try
several things such as adding more training data and increasing the number
of epochs.

model.save_weights('dogs_vs_cats.h5')

# Let's now use our model to classify a few images. dogs=1, cats=0
# Let's start with dogs.
img = cv2.imread("/data/test/dogs/dog.148.jpg")

img = np.array(img).astype('float32')/255
img = cv2.resize(img, (150,150))
plt.imshow(img)
plt.show()
```

```
img = img.reshape(1, 150, 150, 3)
```

```
print(model.predict(img))
[[0.813732]]

print(round(model.predict(img)))
1.0

# Another one
img = cv2.imread("/data/test/dogs/dog.235.jpg")
img = np.array(img).astype('float32')/255
img = cv2.resize(img, (150,150))
plt.imshow(img)
plt.show()
```

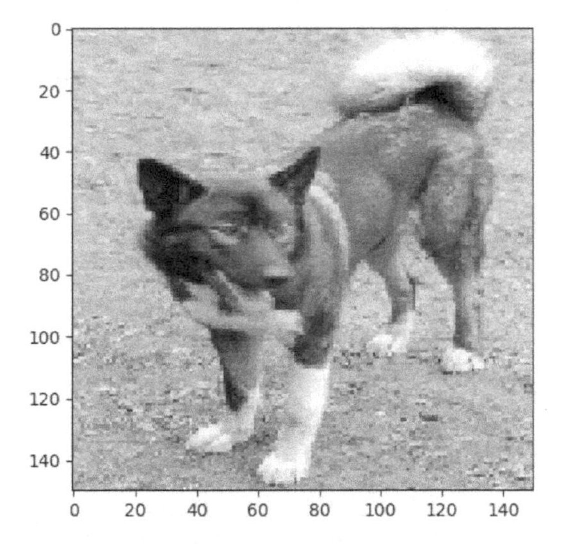


```
img = img.reshape(1, 150, 150, 3)

print(model.predict(img))
[[0.92639965]]

print(round(model.predict(img)))
1.0

# Let's try some cat photos.

img = cv2.imread("/data/test/cats/cat.355.jpg")
```

```
img = np.array(img).astype('float32')/255
img = cv2.resize(img, (150,150))
plt.imshow(img)
plt.show()
```

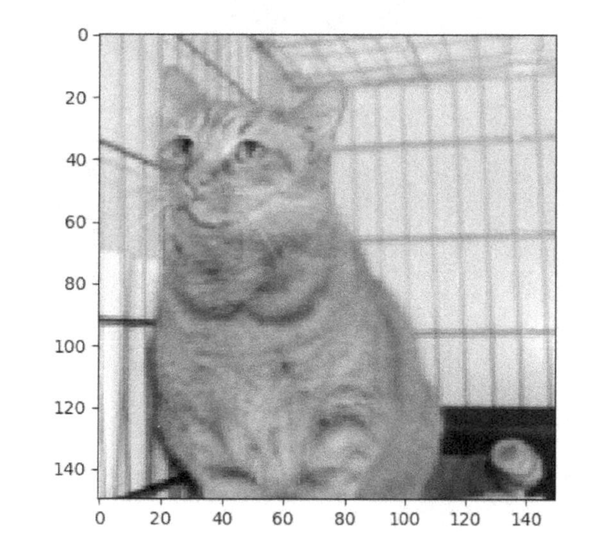


```
img = img.reshape(1, 150, 150, 3)

print(model.predict(img))
[[0.49332634]]

print(round(model.predict(img)))
0.0
# Another one
img = cv2.imread("/data/test/cats/cat.371.jpg")

img = np.array(img).astype('float32')/255
img = cv2.resize(img, (150,150))
plt.imshow(img)
plt.show()
```

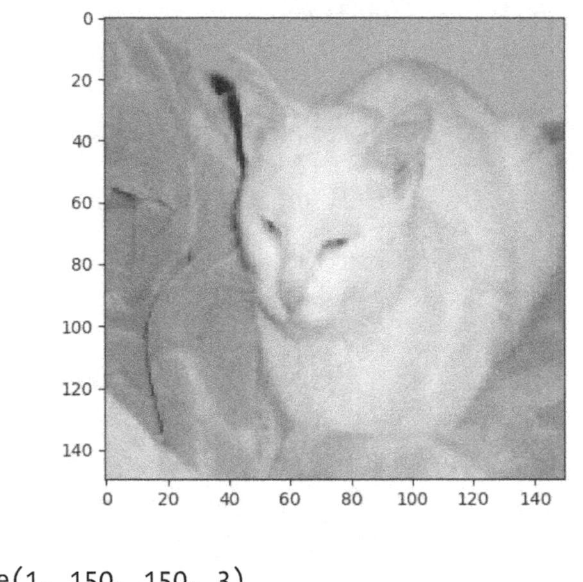


```
img = img.reshape(1, 150, 150, 3)

print(model.predict(img))
[[0.16990553]]

print(round(model.predict(img)))
0.0
```

Summary

This chapter provided you with an introduction to deep learning and distributed deep learning with Spark. I chose Keras for deep learning due to its simplicity, ease of use, and popularity. I chose Elephas and Dist-Keras for distributed deep learning to keep the ease of use of Keras while enabling highly scalable deep learning workloads with Spark. In addition to Elephas and Dist-Keras, I recommend you explore other non-Spark distributed deep learning frameworks such as Horovod. For a more in-depth treatment of deep learning, I recommend *Deep Learning with Python* by Francois Chollet (Manning) and *Deep Learning* by Ian Goodfellow, Yoshua Bengio, and Aaron Courville (MIT Press).

References

i. Francis Crick, The Astonishing Hypothesis: The Scientific Search for the Soul, Scribner, 1995

ii. Michael Copeland; "What's the Difference Between Artificial Intelligence, Machine Learning, and Deep Learning?", nvidia.com, 2016, `https://blogs.nvidia.com/blog/2016/07/29/whats-difference-artificial-intelligence-machine-learning-deep-learning-ai/`

iii. NVIDIA; "Deep Learning," developer.nvidia.com, 2019, `https://developer.nvidia.com/deep-learning`

iv. DeepMind; "Alpha Go," deepmind.com, 2019, `https://deepmind.com/research/case-studies/alphago-the-story-so-far`

v. Tesla; "Future of Driving," tesla.com, 2019, `www.tesla.com/autopilot`

vi. Rob Csongor; "Tesla Raises the Bar for Self-Driving Carmakers," nvidia.com, 2019, `https://blogs.nvidia.com/blog/2019/04/23/tesla-self-driving/`

vii. SAS; "How neural networks work," sas.com, 2019, `www.sas.com/en_us/insights/analytics/neural-networks.html`

viii. H2O; "Deep Learning (Neural Networks)," h2o.ai, 2019, `http://docs.h2o.ai/h2o/latest-stable/h2o-docs/data-science/deep-learning.html`

ix. Michael Marsalli; "McCulloch-Pitts Neurons," ilstu.edu, 2019, `www.mind.ilstu.edu/curriculum/modOverview.php?modGUI=212`

x. Francis Crick; Brain Damage p. 181, Simon & Schuster, 1995, Astonishing Hypothesis: The Scientific Search for the Soul

xi. David B. Fogel; Defining Artificial Intelligence p. 20, Wiley, 2005, Evolutionary Computation: Toward a New Philosophy of Machine Intelligence, Third Edition

xii. Barnabas Poczos; "Introduction to Machine Learning (Lecture Notes) Perceptron," cmu.edu, 2017, www.cs.cmu.edu/~10701/slides/Perceptron_Reading_Material.pdf

xiii. SAS; "History of Neural Networks," sas.com, 2019, www.sas.com/en_us/insights/analytics/neural-networks.html

xiv. Skymind; "A Beginner's Guide to Backpropagation in Neural Networks," skymind.ai, 2019, https://skymind.ai/wiki/backpropagation

xv. Cognilytica; "Are people overly infatuated with Deep Learning, and can it really deliver?", cognilytica.com, 2018, www.cognilytica.com/2018/07/10/are-people-overly-infatuated-with-deep-learning-and-can-it-really-deliver/

xvi. Yann LeCun; "Invariant Recognition: Convolutional Neural Networks," lecun.com, 2004, http://yann.lecun.com/ex/research/index.html

xvii. Kyle Wiggers; "Geoffrey Hinton, Yann LeCun, and Yoshua Bengio named Turing Award winners," venturebeat.com, 2019, https://venturebeat.com/2019/03/27/geoffrey-hinton-yann-lecun-and-yoshua-bengio-honored-with-the-turing-award/

xviii. Alex Krizhevsky, Ilya Sutskever, Geoffrey E. Hinton; "ImageNet Classification with Deep Convolutional Neural Networks," toronto.edu, 2012, www.cs.toronto.edu/~fritz/absps/imagenet.pdf

xix. PyTorch; "Alexnet," pytorch.org, 2019, https://pytorch.org/hub/pytorch_vision_alexnet/

xx. Francois Chollet; "Deep learning for computer vision," 2018, Deep Learning with Python

xxi. Andrej Karpathy; "Convolutional Neural Networks (CNNs / ConvNets)," github.io, 2019, http://cs231n.github.io/convolutional-networks/

xxii. Jayneil Dalal and Sohil Patel; "Image basics," Packt Publishing, 2013, Instant OpenCV Starter

xxiii. MATLAB; "Introducing Deep Learning with MATLAB". mathworks.com, 2019, www.mathworks.com/content/dam/mathworks/tag-team/Objects/d/80879v00_Deep_Learning_ebook.pdf

xxiv. Vincent Dumoulin and Francesco Visin; "A guide to convolution arithmetic for deep Learning," axiv.org, 2018, https://arxiv.org/pdf/1603.07285.pdf

xxv. Anish Singh Walia; "Activation functions and it's types-Which is better?", towardsdatascience.com, 2017, https://towardsdatascience.com/activation-functions-and-its-types-which-is-better-a9a5310cc8f

xxvi. Skymind; "Comparison of AI Frameworks," skymind.ai, 2019, https://skymind.ai/wiki/comparison-frameworks-dl4j-tensorflow-pytorch

xxvii. Nihar Gajare; "A Simple Neural Network in Keras + TensorFlow to classify the Iris Dataset," github.com, 2017, https://gist.github.com/NiharG15/cd8272c9639941cf8f481a7c4478d525

xxviii. BigDL; "What is BigDL," github.com, 2019, https://github.com/intel-analytics/BigDL

xxix. Skymind; "Distributed Deep Learning, Part 1: An Introduction to Distributed Training of Neural Networks," skymind.ai, 2017, https://blog.skymind.ai/distributed-deep-learning-part-1-an-introduction-to-distributed-training-of-neural-networks/

xxx. Skymind; "Distributed Deep Learning, Part 1: An Introduction to Distributed Training of Neural Networks," skymind.ai, 2017, https://blog.skymind.ai/distributed-deep-learning-part-1-an-introduction-to-distributed-training-of-neural-networks/

xxxi. Databricks; "Deep Learning Pipelines for Apache Spark," github.com, 2019, https://github.com/databricks/spark-deep-learning

xxxii. Max Pumperla; "Elephas: Distributed Deep Learning with Keras & Spark," github.com, 2019, https://github.com/maxpumperla/elephas

xxxiii. Max Pumperla; "mnist_lp_spark.py," github.com, 2019, https://github.com/maxpumperla/elephas/blob/master/examples/mnist_mlp_spark.py

xxxiv. Joeri R. Hermans, CERN IT-DB; "Distributed Keras: Distributed Deep Learning with Apache Spark and Keras," Github.com, 2016, https://github.com/JoeriHermans/dist-keras/

xxxv. Microsoft Research; "Dogs vs. Cats," Kaggle.com, 2013, www.kaggle.com/c/dogs-vs-cats/data

xxxvi. Francois Chollet; "Building powerful image classification models using very little data," keras.io, 2016, https://blog.keras.io/building-powerful-image-classification-models-using-very-little-data.html

Index

B. Quinto, *Next-Generation Machine Learning with Spark*, https://doi.org/10.1007/978-1-4842-5669-5

D

E

F

G, H

I

J

K

L

M

N

O

P, Q

R

S

CPSIA information can be obtained
at www.ICGtesting.com
Printed in the USA
LVHW061759260220
648295LV00006B/120

9 781484 256688